Studies in Systems, Decision and Control

Volume 60

Series editor

Janusz Kacprzyk, Polish Academy of Sciences, Warsaw, Poland
e-mail: kacprzyk@ibspan.waw.pl

About this Series

The series "Studies in Systems, Decision and Control" (SSDC) covers both new developments and advances, as well as the state of the art, in the various areas of broadly perceived systems, decision making and control- quickly, up to date and with a high quality. The intent is to cover the theory, applications, and perspectives on the state of the art and future developments relevant to systems, decision making, control, complex processes and related areas, as embedded in the fields of engineering, computer science, physics, economics, social and life sciences, as well as the paradigms and methodologies behind them. The series contains monographs, textbooks, lecture notes and edited volumes in systems, decision making and control spanning the areas of Cyber-Physical Systems, Autonomous Systems, Sensor Networks, Control Systems, Energy Systems, Automotive Systems, Biological Systems, Vehicular Networking and Connected Vehicles, Aerospace Systems, Automation, Manufacturing, Smart Grids, Nonlinear Systems, Power Systems, Robotics, Social Systems, Economic Systems and other. Of particular value to both the contributors and the readership are the short publication timeframe and the world-wide distribution and exposure which enable both a wide and rapid dissemination of research output.

More information about this series at http://www.springer.com/series/13304

Krishna Nath Pandey

Paradigms of Knowledge Management

With Systems Modelling Case Studies

Springer

Krishna Nath Pandey
HRD
Power Grid Corporation of India Limited
Gurgaon, Haryana
India

ISSN 2198-4182 ISSN 2198-4190 (electronic)
Studies in Systems, Decision and Control
ISBN 978-81-322-2783-0 ISBN 978-81-322-2785-4 (eBook)
DOI 10.1007/978-81-322-2785-4

Library of Congress Control Number: 2016936576

Printed on acid-free paper

This Springer imprint is published by Springer Nature
The registered company is Springer (India) Pvt. Ltd.

To
Mr. Ravi Prakesh Singh
for his knowledge, humility and suave leadership

Preface

The knowledge economy, enabled by globalization and virtualization, has found that intellectual capital is the first and foremost requirement of businesses today. Intellectual capital comprises of human capital, social capital, and organizational or structural capital. The management of intellectual capital is an integral task of Knowledge Management. However, the implementation of Knowledge Management in organizations has brought mixed results to the fore. This is probably due to the fact that there are no fixed models available for the practice of Knowledge Management, especially its implementation in organizations. This could also be attributed to the lack of clarity of role of the mechanism of and the factors shaping Knowledge Management. This book is an attempt to address these issues in specific contexts, so that the problems related to Knowledge Management Implementation can be mitigated.

Motivation

This book is aimed at graduate students and professionals involved in Knowledge Management. The key features of the book are as follows:

- It provides the background of Knowledge Management including principles, concepts, models, framework, processes and theories by giving a bird's eye view of various definitions.
- It details Knowledge Management Implementation by delineating the processes, tools and drivers required by an organization, because these are the important aspects of the context specificity of Knowledge Management.
- It highlights the fact that Knowledge Management may be implemented as part of change management and change of culture. Thus underlining that this task can be assigned largely to human resources management enabled by information and communication technology.

- It enables the readers to have confidence in Knowledge Management Implementation and its capitalization.
- It works towards developing enthusiasm in readers towards taking Knowledge Management positively, as it is the need of the hour.
- It provides an in-depth understanding of Knowledge Management to researchers by the microcosm of methodology.

Focus and Target

This book is planned in an exclusive way and is quite different from existing books on Knowledge Management. It covers Knowledge Management comprehensively, from envisioning to evaluation, through a single case study. This method pinpoints the ways and means for successfully implementing Knowledge Management covering all types of knowledge-tacit, explicit and implicit.

The book comprises of six chapters. The first chapter provides a background study covering Knowledge Management in its entirety, whereas the second chapter provides the context through which the hypothesized model is developed. The third chapter provides details of the Company which has been identified for study, data collection and interpretation. The fourth chapter deals with the methodology that has been adopted for study including the softwares used. The penultimate chapter provides details on the significant model after hypotheses testing and the technology acceptance model. The last chapter shows the dominance of the human resource management factors over tenets of information and communication technology in Knowledge Management. The indication for future exploration in the realm of Knowledge Management also finds place in this chapter.

Gurgaon, Haryana, India Krishna Nath Pandey

Acknowledgements

Three great individuals—Mr. I S Jha, CMD POWERGRID; my guru Prof. S. Srilatha, Director, School of Management Studies, IGNOU, and Mr. B.K Banerjee my mentor in NTPC and POWERGID—induced the habit of learning, inferring, and understanding. I am indebted to these three great minds for initiating me in the field of Knowledge Management. Mr. Shwet Kumar Pandey, an expert statistician, did significant hand holding during the navigation of using the statistical tools for this study. Thank you, Shwet.

My colleagues at POWERGRID, Ms. Meenakshi Davar, ED(HR), Mr. Anil Saberwal, GM(HRD), Mr. V.K Singh, DGM(HRD), Mr. Yatindra Dwivedi, DGM (HR), Mr Sachin Kumar, DM(HRD) and Mr. Bhupinder Dhaka, Sr. Engineer (HRD) provided all the support needed for penning down this book. My personal thanks to all these outstanding professionals. Mr Chetan Varma, GM(CC), introduced me to the world of writing and publishing for which I am thankful.

The three persons—my wife Mrs. Mala Pandey, daughters Neha Pandey and Medha Pandey—tolerated my tantrums smilingly and sacrificed their precious time for enabling me to concentrate on my study. A big thank you to all of you! Thanks also to my son-in-law Ganesh who provided constructive suggestions.

Kudos to the editorial team at Springer who helped to improve the manuscript and bring it to a readable shape! Last but not the least, my elder brother Mr. Vashista Muni Pandey and my sister-in-law Mrs Malati Pandey allowed the time needed for studies in my childhood, my reverence to them!

Krishna Nath Pandey

Contents

About the Author

Krishna Nath Pandey is Deputy General Manager in the Human Resource Development Department of the Power Grid Corporation of India Ltd (POWERGRID). He is leading a team of professionals for Knowledge Management and ensuring learning and development of the personnel of his Company. He holds three post graduate degrees in the fields of Management, English Literature and Linguistics. He has topped at the M.A (English Litt) and M.A (Linguistics) examinations. He has also submitted his Ph.D. thesis in Knowledge Management for consideration at IGNOU, India. He has catered to the developmental needs of the work-force of POWERGRID, especially from the perspective of Knowledge Management. Mr. Pandey is a life member of National Institute of Personnel Management. Besides starting his career as a Lecturer in English at Tribhuvan University Nepal, he also served as an officer in Ministry of HRD, Government of India. Afterwards he joined NTPC as an executive in Personnel & Administration. Mr. Pandey has been publishing research papers in national and international journals, in the areas of Knowledge Management, cloud computing and HR. He contributed a chapter on HRM practices for increase of productivity in a book on HR practices. He serves as a visiting faculty member in the area of Knowledge Management at IIM Lucknow, Amity University, MDI, NPTI and other reputed institutions. His areas of interest are Knowledge Management and analytics, human resource development, management control systems, and HR interventions in upcoming companies. He is also authoring a book titled 'Behavioral Aspects of Green Economy'. He has been appointed as a member of Wage Revision Committee of Government of Delhi.

List of Figures

List of Tables

Chapter 1
Introduction

Knowledge Management is a vast area comprising of creation, acquisition, collation, sharing, use, reuse and capitalization of knowledge in an organization. During last three decades, Knowledge Management has been in vogue and a number of companies have tried to implement Knowledge Management for competitive advantage, adoption of best practices, reduction of time and money and avoiding 'reinvention of wheel'. However, Knowledge Management is to be studied from envisioning to evaluation in the context of a Company so that a framework could be given for emulation and practice by other companies. Since Knowledge Management has started 'to become a big focus', the secret of the success of Knowledge Management is also to be underlined.

The Knowledge Management Implementation literature depicts that there are various phases and generations of Knowledge Management. Phases include 'assembly of a realization team, analysis of the initial state, and creation of a knowledge strategy and realization of Knowledge Management activities' (Bures 2009). The three generations of Knowledge Management are first generation, which focused on timely information provisions for decision support and business process re-engineering; second generation, which was triggered by 'SECI model of Nonaka'; and third generation, which is in 'context, narrative and content management' (Snowden 2002). The stages of Knowledge Management have also been brought out as first stage which is field driven by IT, the second stage is seen as focusing on human factors, organizational learning and knowledge creation among tacit and explicit knowledge, whereas third stage shifts to the 'arrangement and management of content through taxonomy, construction and use' (Mvungi and Jay 2010). It may be seen that human resource aspect has been major driving force in Knowledge Management through all the phases, stages and generations. In this work, people-related issues of Knowledge Management Implementation will be studied with reference to POWERGRID.

K.N. Pandey, *Paradigms of Knowledge Management*, Studies in Systems, Decision and Control 60, DOI 10.1007/978-81-322-2785-4_1

1.1 Background and Evolution of Knowledge Management

The universally acknowledged truth is that knowledge economy has pervaded the entire world and 'Indian Knowledge economy has definitely identified knowledge firms as the primary driver' (Richter and Banerjee 2003, p. 2). The massive growth in information technology and knowledge transfer across the boundaries of the organization and eventually the nation has provided a major impetus to the phenomenon of globalization (Cheng 2004; Green and Ruheler 1995). Organizational knowledge sharing has gained recognition against this backdrop as one of the most important keys to performance in organizations. Successful organizations today reward employees for seeking and sharing organizational knowledge, thereby creating a 'knowledge pull'—a grass root desire among employees to tap into their organization's intellectual resources (Hauschild et al. 2001). The onus on organizations to develop and maintain effective knowledge-sharing practices that can transcend organizational and national boundaries is even more given the globalization of businesses today. Apropos this, it attains importance that the entire gamut of Knowledge Management at organizational level is studied in detail from six points of view: (i) theoretical aspects of Knowledge Management; (ii) processes of Knowledge Management; (iii) organizational and social aspects of Knowledge Management; (iv) managerial aspects of Knowledge Management; (v) technological aspects of Knowledge Management; and (vi) application-specific Knowledge Management (Schwartz 2006).

1.2 Theoretical Aspects of Knowledge Management

In this section, an attempt has been made to delineate the strings of various shades of thoughts which prepare the blend of concepts and reify them in a commingling of theoretical depiction of knowledge and its management by various organizations. During the process of browsing the literature for the 'concept of Knowledge Management', 'theories of Knowledge Management' and 'principles of Knowledge Management', it came to the fore that clear-cut demarcation does not exist in literature. However, while delving into the electronic resources and printed material, some clarity emerged. At this juncture, it is better to understand these nuances of the theoretical formulations in a little detail.

1.2.1 Concept of Knowledge Management

In a seminal work, Szulanski and Cappetta (2003, pp. 515–517) talked about the involvement of many actors in the concept of knowledge and they also posit that the concept of knowledge is a 'conception of stickiness'. And if the knowledge is to be transferred from one actor to the other in an organization, 'it proceeds as expected and yields the desired results'. A suggestion regarding the concept of knowledge was also made that 'in a difficult transfer, problems are likely to escalate'. The scholars also underlined that if the concept of knowledge is used for 'successful replication of results', then costly omissions are more likely when there is 'causal ambiguity'. However, the seedlings threw up the features of concept of knowledge and the involvement of 'actors', 'stickiness', 'transferability' and 'causal ambiguity'.

Aungus (2010) further elaborated the inter-relatedness of the concept of Knowledge Management with an absence of common naming concept and features which create problems for technology professionals to provide the enabling technology for Knowledge Management. Nevertheless, he proved that 'Knowledge Management is the concept under which information is turned into actionable knowledge (p. 73)' and made available effortlessly in a usable form to the people who can apply it. It is clear here that a new dimension of information has been brought in the realm of the concept of Knowledge Management. However, this addition of information was an emphasis of the earlier ideas of Graff and Jones (2003) when they were tracing the centrality of 'knowledge and information' as the 'key assets' for organizations. They also pinpointed that three terms—data, information and knowledge—are the parts of the concept of Knowledge Management.

The evolution of the concept of Knowledge Management was taking shape and getting crystallized which is manifested by studies of some of the scholars simultaneously. In the year 2011, Mruthyunjaya (p. 46) talked about 'transformation of many intangibles together with certain tangibles'. He also opined that this should be preceded by the process and termed it as 'the advanced stage of Knowledge Management'. He proposed that the concept of Knowledge Management should usher in 'preparing prototypes or concept models during development activities which typically represent this situation since prototypes primarily serve the purpose of either confirming adequacy of the knowledge that has gone to generating the prototype or its insufficiency, implying that the acquired knowledge is not adequate for the purpose and needs to be strengthened further'.

Baloh et al. (2011) pondered over the issue of concept of Knowledge Management in greater detail and besides bringing under its umbrella the philosophical views including epistemological and ontological perspectives, they concluded that the relation of concept of Knowledge Management permeates 'data, information, intelligence and wisdom'. They also included that 'organizational learning in the gambit of concept of Knowledge Management and showed how it is related to the knowledge possessed by an entity' (p. 36). They also advocated that while dealing with the conceptual aspects of Knowledge Management, economic and social traits of knowledge, dimensions of knowledge along with its volume and volatility should also be analysed.

1.2.1.1 Inter-relatedness of Data, Information, Knowledge and Wisdom

There are two inferences of these terms: one that is the meaning as depicted in dictionary, and the other is as presented by the scholars of Knowledge Management. The dictionary meanings (Soanes et al. 2008) of these terms are given in following lines:

(i) Data:

1. Things given or granted; things known or assumed as facts and made the basis of reasoning or calculation.
2. Facts, especially numerical facts, collected together for reference or information.
3. The quantities, characters or symbols on which operations are performed by computers and other automotive equipment, and which may be stored and transmitted in the form of electrical signal, records or magnetic, optical or mechanical recording media.

(ii) Information:

1. Formation or the moulding of the mind on character, training, instruction, teaching.
2. Communication of the knowledge of some fact on occurrence.
3. Knowledge or facts communicated about a particular subject, event, etc.: intelligence, news.

(iii) Knowledge:

1. The fact of knowing a thing state, person, etc.: acquaintance, familiarity gained by experience.
2. Acquaintance with a fact or facts, a state of being aware or informed: awareness, consciousness.
3. Intellectual perception of fact or truth; clear and certain understanding or awareness, especially as opposed to opinion.
4. The fact or condition of being instructed, or of having information acquired by study or research. Also, a person's range of information learning erudition.

(iv) Wisdom:

1. The quality of being wise, especially in relation to conduct and the choice of means and ends; the combination of experience and knowledge, with the ability to apply them judiciously, sound judgment, prudence, practical sense.
2. Knowledge especially of an abstruse, kind enlightenment, learning erudition.
3. Wise discourse on teaching.

With the above, it is clear that the semantic fields of data, information, knowledge and wisdom are intertwined and overlapping. However, scholars of Knowledge Management have tried to provide a definite connotation to these terms. Some important and representative interpretations have been depicted in Table 1.1.

By looking at the semantic and professional points, it can be concluded that data has got universal value, but it remains meaningless until it is put to some context. Once the data gets contextualized, it attains meaning and metamorphoses itself into information which can be used. The apt and appropriate use of information at the right moment may be termed as knowledge. These inferences are strictly deduced apropos managerial repertoire. With this perspective, wisdom can be seen as that faculty of mind which ensures maximum returns from the probabilities.

1.2.2 Learning Organization

An important part of the concept of Knowledge Management is the 'learning organization'. In fact, the business world is not the same anymore because knowledge and information are being increasingly described as resources, powerbases, assets, competitive advantage, strategic weapons and so on (Shukla 1997). Corporate learning as a leverage for organizational transformation requires the designing of processes to ensure regular flow of relevant information across hierarchical, functional and corporate boundaries, and the creation of systems facilitates the constructive use of this information for organizational problem-solving at all levels.

Thus, a learning organization is one which includes the following:

- The procedures by which knowledge and practice are transmitted across posting cycles, across different work situations and across time;
- 'The procedures that facilitate generative learning—learning that enhances the enterprise's ability to adjust to dynamic and unexpected situations and to react creatively to them'. (Ali et al. 2006, p. 561) Such a situation entails that the 'transdisciplinary' approaches are taken by the organizations which is pebbled with 'heterogeneity of skills' and it should necessarily be context-sensitive, where the producers and users of the knowledge are interacting with each other quite often.

Scholars like Eisenhardt and Galunic (2001) had used the term 'coevolving', which is crucial for knowledge intensive corporations, and they opined that for learning organizations it is a subtle strategic process (p. 132). Such a situation warrants that managers need employees who think constantly and creatively about the needs of the organization. Only such collaboration will lead to the 'learning organization'. To conclude the discussion, it may be said that a learning organization keeps scanning the business environment, may be through 'semantic clustering analysis', 'stakeholder analysis' and lists its knowledge requirements regularly to update its Knowledge Management strategy, knowledge assets and intellectual capital.

Table 1.1 Inter-relationship of data, information, knowledge and wisdom

S. No.	Author and year	Data	Information	Knowledge	Wisdom
1	Davenport and Prusak (2000)	'Data is a set of discrete, objective facts about events' (p. 2)	'Information is a message usually in the form of a document or an audible or visible communication' (p. 3)	'Knowledge is a fluid mix of framed experience, values, contextual information and expert insight that provides a framework for evaluating and incorporating new experiences and information. It originates and is applied in the minds of knowers. In organizations, it often becomes embedded not only in documents and repositories but also in organizational routines processes, practices, and norms' (p. 5)	–
2	Tiwana (2000a)	'Raw transactional representations and outputs without inherent meaning' (p. 590)	–	'Knowledge is simply actionable information' (p. 57)	–
3	O'Dell (2004)	'Facts and figures presented out of context for the purpose of innovation or informed efficiency' (p. 121)	'Data that is presented in context so people might make use of it' (p. 123)	'Information in action; information that people make use of along with the rules and context of its use' (p. 123)	–

(continued)

Table 1.1 (continued)

S. No.	Author and year	Data	Information	Knowledge	Wisdom
4	Debowski (2006)	–	'*…access to libraries and other sources to find out best practice.' (p. 330)	'The process of translating information and past experience into a meaningful set of relationships which are understood and applied by a person' (p. 348)	–
5	Vedpuriswar (2009)	'A set of particular and objective facts about an event or a transaction' (p. 83)	'Information is processed data. Data becomes information when it is summarized tabulated processed and checked for errors'. (p. 105)	'Understanding, clarity and insights that we gain through education, practical experience, reflection and observing others' (p. 111)	'Understanding clearly which knowledge to use for what purpose and wisdom is the ability to make correct judgments and decisions' (p. 176)
6	Warrier (2009)	'Raw data is gathered from nominated, grated, operational and legacy applications, cleansed and then summarized and presented in a way that makes sense to end users' (p. 143)	'Information is generated by organizing data within contexts' (p. 11)	'A useful definition of knowledge relevant to the context is that when knowledge is acted upon it provides meaning for information and data and directs decision making'. (p. 96)	'Wisdom is the state of the human mind characterized by profound understanding and deep insight' (p. 89)
7	Hawryszkiewycz (2010)	–	'Primarily information is something that is recorded and readily available whereas knowledge is created by interpreting this information' (p. 16)	'Knowledge is something that is abstract in reality, it is hard to point to something and say there is knowledge.' (p. 72)	–
8					

(continued)

Table 1.1 (continued)

S. No.	Author and year	Data	Information	Knowledge	Wisdom
	Desouza and Pequette (2011)	'Raw facts and numbers that are not useful or meaningful on their own, and require interpretation for processing in order to add value to the organization' (p. 330)	'Data that has been converted to have meaning in a particular context' (p. 332)	The belief of an individual based on the meaningful accumulation of information. It can be derived from experiences, or through the identification and use of other information and knowledge resources' (p. 333)	'An individual's accumulated knowledge and experience applied to a particular context.' (p. 336)
9	Pasher and Ronen (2011)	'Data is raw material of Knowledge Management' (p. 187)	'When you process this raw material (data) and today it is possible to process it using various technological systems it becomes information. Information is processed data' (p. 188)	'To actually solve a problem backed up by data, you need to go up a level. We call this level, above data and information, knowledge' (p. 188)	'…it is *wisdom* (the ability to identify which knowledge has the potential to be intellectual capital) that is worth investing in and developing because there are good chances that it will generate future's most desired outcome' (p. 189)

1.2.3 *Taxonomy of Knowledge*

While delving over the aspect of concept of Knowledge Management, it came to the fore that a scholar, namely Ein-Dor (2006), has proposed the taxonomies of knowledge. He proposed five major dimensions and one additional dimension, namely the tacit–explicit dimension, individual–social dimension, procedural–declarative dimension, commonsense–expert dimension, and task–context dimension, and, of course, the additional dimension includes true–false, certain–uncertain, private and individual, and public and shared. An attempt was made to delineate this aspect of concept of Knowledge Management. During this course, we came across a book 'Knowledge Management: Participants Guide' published by Asian Productivity Organization (2012) which is basically the Knowledge Management module for the participants to a workshop of Asian Productivity Organization, where word taxonomy is not used instead 'types of knowledge' has been used (p. 51). While perusing the literature available under this head, it was noticed that an additional type of knowledge, i.e. 'implicit knowledge', is in circulation and may soon attract the attention of scholars to probe it further before it is brought out of 'grey literature' to the literature of taxonomy of knowledge.

1.3 Principles of Knowledge Management

Principle is a truth or general law that is used as a basis for a theory or a system of belief. Principles of Knowledge Management are no different. It entails that the 'fudgy' ideas find a way to 'express the inexpressible. Unfortunately, one of the most powerful management tools for doing so is also among the most frequently overlooked: the store of figurative language and symbolism that managers can draw from to articulate their intuitions and insights' (Nonaka 1998, p. 31). Taking a cue from this, it was found pertinent to understand that the principles of Knowledge Management should be seen as an outcome of Knowledge Management evolution which is a dynamic process of strengthening organizational effectiveness by maximizing the knowledge utilization (Lin 2011).

One scholar, namely Vyas (2000), compiled a list of Knowledge Management principles after painstaking effort having reviewed the literature and underlining the major contribution from Davenport which are culled out and enumerated below:

(i) Knowledge originates and resides in people's minds.
(ii) Knowledge sharing requires trust.
(iii) Technology enables new knowledge behaviours.
(iv) Knowledge sharing must be encouraged and rewarded.
(v) Management support and resources are essential for Knowledge Management.

(vi) Knowledge initiatives should begin with a Pilot programme.
(vii) Quantitative and qualitative measurements are needed to evaluate the initiative.
(viii) Knowledge is creative and should be encouraged to develop in unexpected ways (p. 27).

The literature is permeated largely with the focus on knowledge as a 'justified personal belief that increases an individual's capacity to take effective action' (Huber 1991, p. 88).

The evolution in encapsulation of Knowledge Management principles drew an analogy of Toyota's 'lean' principles to the realm of Knowledge Management principles also by scholars, viz., Staats and Upton (2011) who reported their research apropos Knowledge Management principles on following six counts:

(a) Continually rooting out waste should be an integral part of every knowledge worker's job.
(b) Strive to make tacit knowledge explicit.
(c) Specify how workers should communicate with one another.
(d) Use the scientific method to solve problems as soon as possible. The people who created the problem should fix it.
(e) Plan for an incremental journey—start, small, codify the lessons learned, and keep looking for new ways to work.
(f) Leaders must blaze the trail (pp. 86–96).

These principles are limiting the dynamics of Knowledge Management which have evolved from the information-processing view to the sense-making ones. 'The sense-making view of Knowledge Management is not a process of mechanically assembling pieces of knowledge, but a process of organically associating a diverse set of knowledge which results in the production of new developments. Its process should be dynamic, evolving and organic; rather than static, predefined and mechanistic' (Yoo 2011, p. 157)

1.4 Theories of Knowledge Management

Theories are reasoned sets of ideas that are intended to explain why things happen or exist. It entails that the theoretical foundations of Knowledge Management are studied at length. This becomes furthermore important on organizational level. However, the growth and sharing of knowledge is recognized as one of the most important elements in becoming a learning organization (Easterby-Smith 1997; Marsick and Watkins 1994; Senge 1990). What has been missing according to many researchers and practitioners in the field is the development of a theoretical foundation for describing how people learn and perform in an organization (Raybould 1995; Salisbury 2001). This theoretical foundation is needed by today's organizations to avoid the development of technological solutions that do not support their entire Knowledge Management cycle (Plass et al. 2000; Salisbury 2003, p. 131). In order to have an in-depth understanding of the topic, various theories have been traced in the realm of Knowledge Management as put forth by scholars.

1.4.1 Epistemological Theory

Epistemology is 'the branch of philosophy that deals with the varieties, grounds and validity of knowledge' (Shorter Oxford English Dictionary 2007, p. 851). Aaron (2006) studied the relevance of Knowledge Management with the epistemology. He writes that 'as far as Knowledge Management is concerned, there are significant limitations in traditional approaches to epistemology. Traditional epistemology is not concerned with the production and processing of knowledge in group or shared sense—it is not really concerned with the *pragmatics* of knowledge production and use. The main issue in epistemology is the status of the final product rather than the process of getting there and what happens after the knowledge is acquired' (p. 167). However, the compatibility of epistemology has also been traced because though the two disciplines have got different connotations, their base is the same (Aaron 2006). Factual, tacit, practical, technical and other forms of knowledge must still meet certain criteria in order to be genuine knowledge: they must correspond to some aspect of the world, accurately reflect a reliable way of manipulating the world and stand up to the 'harshest of pragmatic test'. The research by Aaron (2006) concluded that, 'The conception of knowledge as developed in recent philosophy of science and epistemology is of great relevance to the pursuit of Knowledge Management. In particular, it seems that these areas could be very useful for developing a theory of collaborative knowledge work' (p. 169).

Frame (2010) further mooted the proposition that 'epistemology is synonymous with theories of knowledge' (p. 1). In fact, epistemology is subdivided into *rationalism* and *empiricism* which are considered as the ways of acquiring knowledge. This theme was further elaborated by taking into account the second look at epistemology. A second important issue in epistemology concerns the ultimate source of knowledge. There are true traditions: empiricism which holds that our knowledge is primarily based on experience and rationalism which holds that our knowledge is primarily based on reason. Although the modern scientific world view borrows heavily from empiricism, there are reasons for thinking that a synthesis of the two traditions is more plausible than either of them individually (Frame 2012). Epistemology also deals with the 'analysis of what is meant by the term "knowledge" itself and with question about the limits and scope of knowledge, its reliability and what constitutes justification for holding a knowledge' (Edgar and Sedgwick 2003).

Through this discussion, it is inferred that the knowledge has its own distinct and independent existence in the societal milieu. This was reaffirmed by a research conducted by Fuller (2005) where he states that 'sociological accounts of knowledge if they have any grounding at all, are grounded in the features of *particular* societies and hence are, in principle opposed to the philosophical accounts which are based on appeals to *universal* rationality' (p. 40). He also indicated that epistemology has been a 'well-motivated autonomous field of enquiry' (p. 41) only in so far as it has been concerned with the social organization of knowledge.

1.4.2 Ontological Theory

Ontology stands for that branch of philosophy which is concerned with the nature of being. Shorter Oxford English Dictionary (2007) defines ontology as 'The science or study of being; that part of metaphysics which relates to the nature or essence of being and existence' (p. 2005). In his seminal work, Carnap (2005) concluded that 'ontological theory does not simply embracing a Platonic ontology but is perfectly compatible with empiricism and strictly scientific thinking' (p. 268). It entails that the ontological theory of Knowledge Management is not only scientifically well-grounded but also delineates as a typical theory in this field. Ontology is an explicit specification of a simplified abstract view of some domain (in this case Knowledge Management) that is to be described, discussed and studied (Gruber 1995).

'The ontology identifies and characterizes basic components of Knowledge Management episodes, the knowledge resources an organization uses in these episodes, a generic set of elemental knowledge manipulation activities that manifest within Knowledge Management episodes, and categories of influence on the conduct and outcome of these episodes' (Holsapple and Joshi 2003, p. 89). In fact, ontology can serve as a common language for discourse about Knowledge Management.

The ontological theory has been used by organizations to the areas of Knowledge Management episodes and the conduct of Knowledge Management, its architecture, required resources, content knowledge resources, knowledge artifacts and all aspects of knowledge including acquisition, creation, sharing, use, reuse, dissemination and capitalization (Holsapple and Joshi 2003). With this spectrum, it can safely be concluded that the ontological theory can be used for the sharing and reuse of knowledge about a domain, providing a foundation for structuring explorations of that domain and enabling advances in our understanding of it. 'Many organizations have adopted ontology development as part of their Knowledge Management initiatives. In such cases, the domain of interest could be the organization's set of knowledge assets and its ways of using those assets: a view of this domain is constructed or identified; this view is explicitly specified as an ontology of concepts and relationships' (Holsapple and Joshi 2003, p. 90).

The muddling of issues of ontology (the study of being—essentially studying questions of what kinds of entities exists) and issues of epistemology (The study of knowing—essentially studying what knowledge is and how it is possible) has been one of the key confusions in philosophy. This has been the case with numerous general schools of philosophy, almost always taking the form of ignoring ontology in favour of epistemology. However, in the subject of Knowledge Management, these two theories are equally useful.

1.4.3 Hermeneutic Theory

Hermeneutic Theory is 'the branch of knowledge that deals with (theory of interpretation)' (Shorter Oxford English Dictionary 2007, p. 1243). Knowledge Management has got many facets which include people components, technology components and process components and because of this, it generates 'a cross-disciplinary domain' (Geisler and Wickramasinghe 2009, p. 16). The people components are subjective, and therefore, an interpretative facet gets added to Knowledge Management especially in the process of knowledge creation by its qualification combination, socialization, externalization, internalization and introspection, thus allowing the transformation of individual experience and knowledge into formalized shareable domain knowledge.

They have offered the findings in the realm of text, historicity and context, the horizons of understanding and the fusion of horizons. They have also pointed out that the hermeneutic circle denotes the global context and an improved understanding of each part which leads to cyclical interpretations. The authors have cautioned that all the 'perceptions and presuppositions' of individuals 'should be acknowledged and incorporated' while documenting the 'observed phenomenon'.

1.4.4 Heuristic Theory

Shorter Oxford English Dictionary (2007) defines heuristic as 'method for attempting the solution of a problem: a rule or item of information used in such a process' (p. 1244). In the context of Knowledge Management, this theoretical formulation is used for new concepts that provide a rationale for Knowledge Management (Baskerville and Dulipovici 2006). Heuristic theory has been depicted as a 'new way of thinking about and pursuing concerns about a foray into many different disciplines, among these are sociology, education, business, strategic decision making, Knowledge Management and organizational theory' (Khin and Fatt 2010, p. 1). A fundamental premise of any applied science is that action in pursuit of one's ends is well served by forethought: it pays to think before you act. Heuristic 'theory of problem structure' (Smith 1988) suggests that alternatives exist or will be identified. Guo and Sheffield (2007) reported that heuristic theory is for exploring the individual choice (diversity) and collective coherence (integration) in Knowledge Management research.

It has also been reported that heuristic-based approach can be used for indexing and clustering of high-dimensional data in the realm of Knowledge Management (Chen et al. 2011). Liu et al. (2012) used heuristic theory for 'evidence' also and concluded that it can be very 'scientific method' in identifying the evidence for knowledge. This discussion leads us to two conclusions: one is that heuristic theory can be used for describing an approach to learning by trying without necessarily having an organized hypothesis or way of proving that the results proved or disproved the hypothesis, and two is pertaining to the use of the general knowledge gained by experience sometimes expressed as 'using a rule-of-thumb'. 'However, heuristic knowledge can be applied to complex as well as simple everyday problems' (?whatis.com 2012, p. 1).

1.4.5 Unified Theory of Knowledge: Attempt for Global Theory

The browsing in electronically available literature by using the terms related with unified theories of Knowledge Management such points as 'universal theory of Knowledge Management', 'Theory of Knowledge Management in Globally distributed setting' and 'unified theory of knowledge' came to the fore which are indicating that scholars are trying to establish a unified theory of knowledge. It is largely due to the fact that 'Knowledge Management has ended up very much like the English language, borrowing vocabularies, concepts, models and approaches from other disciplines. Unlike the English language, it has not built a distinctive identity and literature of its own. There is, in Knowledge Management, very little sense of overall coherence, integration or common agenda has been found. The consequence is a 'chronic sense of malaise, uncertainty and confusion in the field' (Lambe 2011, p. 190). However, this scholar undertook further research and concluded that 'by 2008, Knowledge Management was in the 40 % adoption range, much higher than several other, more recent management tools' (p. 192).

The evolution of Knowledge Management may be traced to very many other disciplines which are antecedents of the repertoire of Knowledge Management literature. 'If the study of Knowledge Management is to have an enduring future, it must take a more dynamic and contextual approach, recognizing that its antecedents come from many more disciplines than those which are cited in literature' (Land et al. 2006, p. 859). It clearly showed that a universal and unified theory was the need of the hour. Nonaka et al. (2001) had been also pestering for such a theory. They exhorted the researchers and scholars because 'the organization is viewed as information—processing machine that takes and processes information in order to solve a problem and adapt to the environment' (p. 491).

Macpherson (2007) tried to provide a 'unified theory of knowledge' when he wrote that 'often the chain of (electrical) signals involves a small number of neutrons …complex brains generally have sensory organs that do some 'pre-processing' of sensory input, but most actively occurs in the brain itself via many neuronal processing executing simultaneously' (p. 13). It is clear from this theory that it does not deal even at tacit level of knowledge. At the most, it is trying to describe the functioning of the brain and that too at the physiological level.

A group of two researchers, namely Pawlowski and Bick (2012), undertook a project towards the development of holistic theory through a 'framework as a step towards building a theory for global Knowledge Management' (p. 93). They had built up their work on the framework developed by CEN (2004) which is 'currently in practice' (pp. 352–364) in the European Standardization Community. They were aware that a global environment structure should have been preceded by 'an analysis structure' (Heisig 2009) covering people issues, organizational structure and processes, required technology and managerial concepts. Pawlowski and Bick (2012) first of all identified 'commonalities of the diverse frameworks (strategies, processes, knowledge resources, tools) and harmonized the different terminologies' (p. 96). While propounding their 'universal theory', they introduced the Global Knowledge Management Framework by including components and influence factors of Knowledge Management in globally distributed settings. The framework also identified the key aspects for designing Knowledge Management Processes and provided a solution space and success factors for decision makers as well as implementers. Once their Global Knowledge Management Framework was ready, they wanted validation for it. With this in view, they took the help of a group of international students to use Global Knowledge Management Framework as an assignment 'to structure their group-work as suggestion' (p. 105). In nutshell, the scholars used the case study research method for testing Global Knowledge Management Framework.

After the assignment was over, Pawloski and Bick found out that Global Knowledge Management Framework 'created a solution space for Global Knowledge Management by providing the Global Knowledge Management Framework identifying and harmonizing Knowledge Management research efforts in the global context'. Based on two scenarios and one proof-of-concept case study (p. 105), this artifact management could be treated as a 'holistic theory' of Knowledge Management which could be universally applicable.

1.5 Definitions of Knowledge Management

At this juncture, a set of Knowledge Management definitions should be perused in order to understand the scope, focus, meaning and nuances of Knowledge Management. Table 1.2 summarizes these aspects of Knowledge Management.

Table 1.2 KM Definitions

S. No.	Author and year	Definition of Knowledge Management
1	Nonaka (1998)	'New knowledge always begins with the individual. A brilliant researcher has an insight that leads to a new patent. A middle manager's intuitive sense of market trend becomes the catalyst for an important new product concept. A shop—floor worker draws on years of experience to come up with a new process innovation. In each case, an individual's personal knowledge is transformed into organizational knowledge valuable to the Company as a whole' (p. 26)
2	Davenport and Prusak (1998)	'Knowledge Management should become part of everything an organization does, and be part of everyone's job. If companies are successful in managing knowledge, they may even forget that they are doing it' (p. XV)
3	Bukowitz and Williams (1999)	'Knowledge Management is the process by which the organization generates wealth from its intellectual or knowledge-based assets' (p. 2)
4	Tiwana (2000a)	'…let's try getting temporary handle on what Knowledge Management means. In the simplest terms it means exactly that: management of knowledge. In the context of our discussion, it can be extended to management of organizational knowledge for creating business value and generating a competitive advantage. Knowledge Management enables the creation, communication and application of knowledge of all kinds to achieve business goals' (p. 5)
5	Alavi and Leidner (2002)	'Since knowledge is personalized, for one person's knowledge to be useful for another individual, it must be communicated in such a manner as to be interpretable and accessible to the other individual. Hoards of information are of little value: only that information which is actively processed in the mind of an individual through a process of reflection, enlightenment and learning can be useful. Knowledge Management, then refers to a systemic and organizationally specified process for acquiring, organizing and communicating both tacit and explicit knowledge of employees so that other employees may make use of it to be more effective and productive in their work' (p. 17)
6	Conway and Sligar (2002)	'Knowledge Management (KM) is not a new term. Over the past decade it has become a generic phrase used to refer to many types of information exchange between people … however, Knowledge Management has a specific meaning: the process of revealing and mapping the work activities, behaviours and knowledge sources within an organization' (p. 1)
7	O' Dell (2004)	'Systematic approaches to help information and knowledge flow to the right people at the right time so they can act more efficiently and effectively. Find, understand, share and use knowledge to create value' (p. 124)

(continued)

Table 1.2 (continued)

S. No.	Author and year	Definition of Knowledge Management
8	Brelade and Harman (2006)	'Common to the various definitions were the two dimensions of people and information. Knowledge Management lies in the relationship of these two dimensions, mediated by systems and processes. It is generally seen as holistic approach which recognizes the inter-dependency for the organizational effectiveness of people, the technology they use and the systems and processes within which they use it' (p. 5)
9	Awad and Ghaziri (2007)	'Knowledge Management (KM) is a newly emerging, interdisciplinary business model that has knowledge within the framework of an organization as its focus. It is rooted in many disciplines, including business, economics, psychology and information management. It is the ultimate competitive advantage for today's firm. Knowledge Management involves people, technology and processes in overlapping parts' (p. 26)
10	Benbya (2008)	'We can trace the roots of KM in several social and economic trends, but in particular associate its emergence with several challenges facing organizations in the twenty-first century. …several features seem to be the basis of the current, renewed importance given to knowledge and its management; we can summarize them as follows: • The proliferation of information and communication technologies and digital networks • The globalization of the economy • The qualitatively different assumption governing knowledge assets • The challenges of lost knowledge • The rise of a service economy' (pp. 13–14)
11	Geisler and Wickramasinghe (2009)	'The business world is increasingly competitive, and the demand for innovative products and services is enormous. In this century of creativity and ideas, the most valuable resources available to any organizations are human skills, expertise and relationships. Knowledge Management (KM) is about capitalizing on these precious assets in a systematic fashion' (p. 3)
12	Hawryszkiewycz (2010)	'Knowledge, however is more difficult to define—one is to see it as using information to make decisions. Knowledge is more on knowing how to interpret information and providing new insights to some problems at hand. It is often using previous experience to interpret the information and use the interpretation to initiate some action. …it is also the ability to transfer knowledge and experience from one context to another' (pp. 72–73)
13	Desouza and Paquette (2011)	'Definitions of Knowledge Management are as varied and myriad as the backgrounds of the authors who propose them. Knowledge Management is a highly interdisciplinary field that attracts scholars and practitioners from such varied areas

(continued)

Table 1.2 (continued)

S. No.	Author and year	Definition of Knowledge Management
		as economics, management, philosophy, innovation, public-Policy, information science, information systems, engineering and sociology among others. With many disciplines contributing to Knowledge Management, a search engine might produce several hundred definition of what is (and is not) Knowledge Management. …Knowledge Management is about increasing an organisation's effectiveness through application of its knowledge assets. It is important to note that Knowledge Management is not exclusive to any particular type of organization. Organisations of all types must manage their knowledge if they are to adequately perform in their environments. No matter what their missions, all organizations have knowledge assets that need to be leveraged' (pp. 4–5)
14	Pasher and Ronen (2011)	'To actually solve a problem backed up by data, you need to go up a level. We call this level, above data and information, *knowledge*. The existence or lack of knowledge is put to the test only through action. When there is a problem, there are those who know to solve it and those who do not know how to solve it. They either have the knowledge or they don't Of course, those who have the knowledge need information to solve the problem, but the information is not enough. They need more. Solving the problem requires experience and expertise, which, when joined with the data and information, becomes knowledge' (p. 188)
15	Johnsen (2012)	'Overall, Knowledge Management is an activity restraint that promotes a shared and integrative move toward the establishment, captive, organization, admission and use of information resources, including the inferred, un-captured knowledge of people. It is the course of action of converting information and rational assets into long-term significance. In an organizational setting this would mean a methodical move toward getting a business to make the best possible use of knowledge in implementing its undertaking, roughly viewed as either sustainable competitive improvement or lasting high routine' (p. 3)
16	Durmusoglu et al. (2014)	'The growing use of knowledge in business contributed to the emergence of the theory of Knowledge Management. Knowledge Management activities can be conceptualized as actuators stimulating the development of new knowledge for the purpose of achieving objectives through identifying, capturing, reusing and leveraging pertinent knowledge. The purpose of Knowledge Management activities in organizations is to ensure growth and continuity of performance by protecting critical knowledge at all levels, applying existing knowledge in all circumstances, combining knowledge in synergistic ways, acquiring relevant knowledge continuously and developing new knowledge through continuous learning' (p. 21)

1.5.1 *Commentary on Definitions of Knowledge Management*

All the definitions of Knowledge Management enumerated in Table 1.2 depict that no final or firm definition has crystallized as yet. However, the gradation of inter-relatedness of data, information and knowledge has once again come to the fore where knowledge is coming into being only on the information which can be used. Nevertheless, the definitions provide a window to the reuse of individual and organizational knowledge by the people who need them in time.

The definitions do not provide any road map as how Knowledge Management can be implemented by organizations. The systems and softwares, if any, have also not been suggested. The main theme of the definitions has clearly shown that it is the need of the hour because of virtualization of organizations, the globalization and arrival of knowledge economy, emergence of service sector and the criticality of creation, acquisition, collation, updation, dissemination, use, reuse, sharing and capitalization of knowledge assets. Finally, it can be concluded that an all pervasive definition of Knowledge Management is yet to arrive on the scene largely because of KM's emergence through interdisciplinary antecedents.

1.6 Knowledge Management Model

1.6.1 *SECI Model*

Models entail that a three-dimensional view may be had of a thing typically on a smaller scale. Knowledge Management models are also given by scholars depicting various dimensions of either creating, conceptualizing, acquiring, sharing, processes, types, practices, stages, flow, use, reuse, robustness, maturity, etc. 'Nonaka and his colleagues' model, in particular the SECI matrix of knowledge conversion, is increasingly being cited by authors in a widening set of disciplines and has evidently achieved something like a paradigmatic status' (Gourlay 2012). This model is naturally to be looked into in a little detail. The model is shown as Fig. 1.1.

The model has been adopted from Nonaka and Takeuchi (1995, pp. 57–71). It can be seen from the model that when it comes to sharing of tacit form of knowledge or its creation from tacit to tacit 'socialization' is the tool. If this involves tacit to explicit, then 'externalization' becomes handy. However, 'combination' is required if it is explicit to explicit and in the case of explicit to tacit, 'internalization' comes into picture. Probably this was the first model in the Knowledge Management and scholars, no doubt, started referring to it while presenting models in other areas of knowledge or its related fields.

McGinnis and Huang (2004, pp. 614–616) used the SECI model while proposing the incorporation of this Knowledge Management model in Enterprise Resource Planning implementation and concluded that the 'spiral' SECI model may now finally be revised. Both of the models proposed by these scholars are depicted in Figs. 1.2 and 1.3.

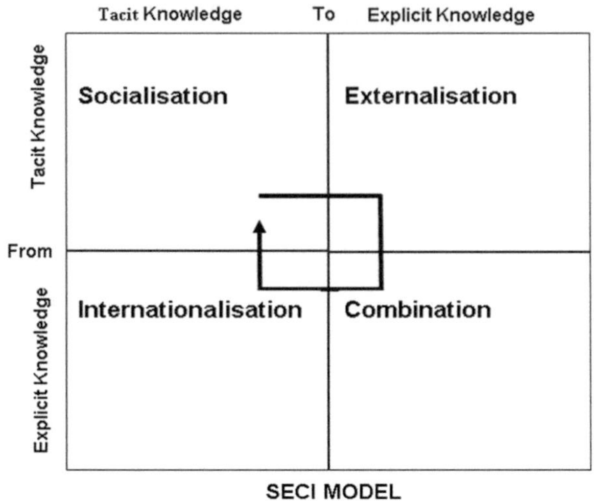

Fig. 1.1 SECI model

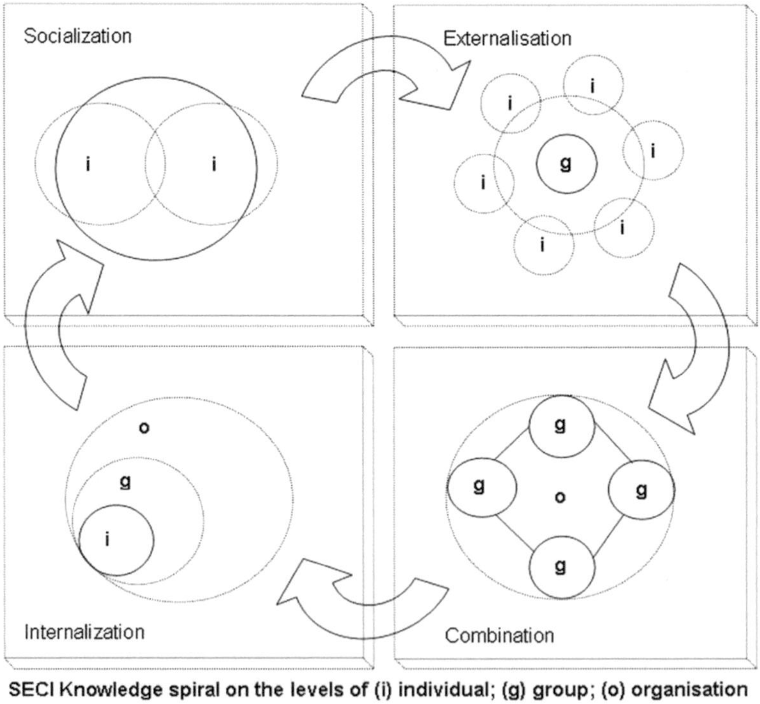

Fig. 1.2 SECI spiral on three levels

Fig. 1.3 Spiral model

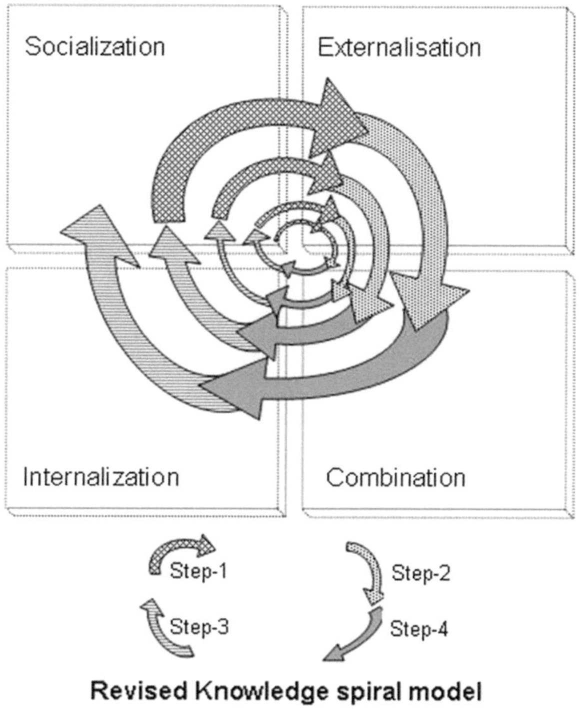

Figure 1.2 of SECI model depicts that the knowledge can have an interplay at three levels: individual (i), group (g) and organization (o). Figure 1.3 finally has been proposed by the scholars as the revised knowledge spiral models were a significant departure from Nonaka et al.'s model of the explicit identification of activities that occur during the progression between quadrants. Nonaka et al.'s model only explained the relationship of knowledge as it is changed from one form to the next. This modified spiral model identifies what activities have to occur within the project phase to support knowledge progression between quadrants.

1.6.2 Hierarchical Model

At this juncture, we intend to present a study by Prat (2006) for his hierarchical model for Knowledge Management. In fact, he presented a summary of the Knowledge Management models which were propounded by various scholars before him and offered a commentary also. He enumerated the KM models offered by scholars (Alavi and Leidner 2001; Davenport and Prusak 1998; Despres and Chauvel 2000; Fowler 2000; Grover and Davenport 2001; Handzic 2001; Holsapple and Joshi 2001; Newman and Conard 2000; Nissen 2002). However, he

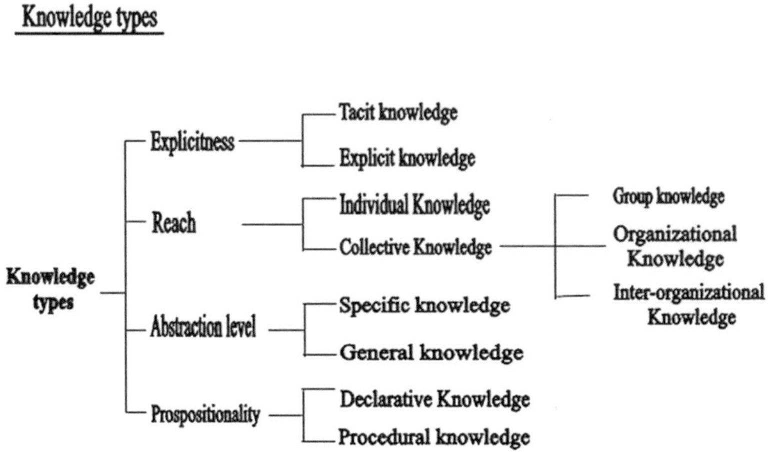

Fig. 1.4 Knowledge types

did not take into consideration the models proposed by Pedersen and Larsen (2001), Hernandez and Sarrano (2012), Zhuge (2002), Malone (2002) nevertheless his summary and commentary holds good when he narrates that 'the existing KM models are often descriptive, abstract, semantic models, that is frameworks. All the models are semantic. Consequently, elaborating on the contributions of these models, our objective is to define an integrated, analytic KM model' (p. 212). He christened his model 'A Hierarchical KM Model'. This model comprises of (i) knowledge types; (ii) KM processes; and (iii) KM context. Since he used these three components, he stipulates that this is in a hierarchy. The three components of hierarchical model are reproduced in Figs. 1.4, 1.5 and 1.6.

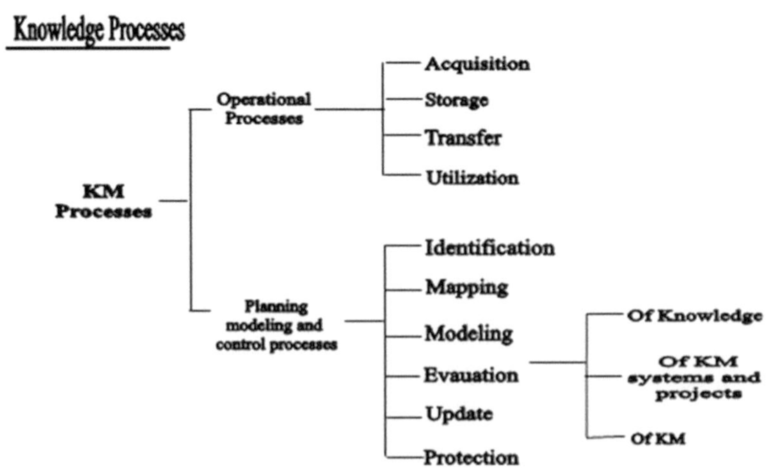

Fig. 1.5 Knowledge processes

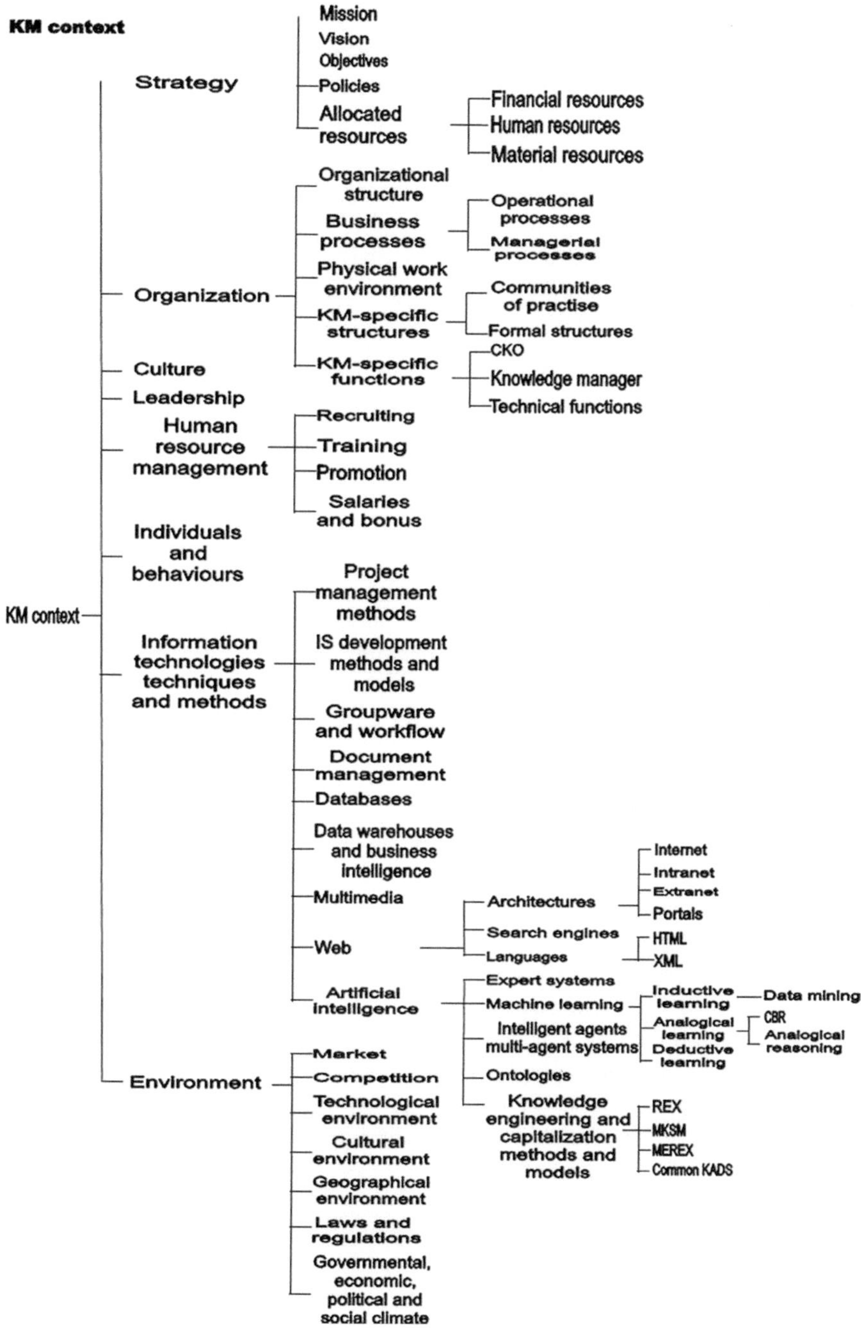

Fig. 1.6 KM context

Prat (2006) has claimed that base on the hierarchical structure of the KM model, the analytic hierarchy process as proposed by other scholar (Saaty 1980) may be applied to the quantitative assessment of KM by taking the following steps:

1. A sub-hierarchy of the hierarchical KM model is selected depending on what needs to be assessed and for what purpose.
2. Weights are determined for the nodes of the sub-hierarchy.
3. The weighed sub-hierarchy is used for performing evaluations. For each evaluated item, scores are entered for the end nodes of sub-hierarchy. Aggregated scores, including the global score of the item are calculated based on the previously defined weights' (pp. 216–217).

This model introduced the possibility of using quantitative assessment of knowledge for the first time in the KM literature.

1.6.3 Process Models

The process stands for a series of actions or steps taken towards achieving a particular end. It also refers to a natural series of change. In the realm of Knowledge Management, the scholars have underlined the existence of knowledge in the process itself. At this juncture, it will not be out of place to look into the process model. In fact, there are two representative process models. Both of them are mentioned below.

1.6.3.1 Process Model of Establishing Knowledge Management

Kjaergaard and Kautz (2006) conducted an empirical study and findings were reported as 'A Process model of establishing Knowledge Management: Insights from a longitudinal filed study' which 'emphasizes how the organizational members make sense of the action and behavior of management and how this understanding influences their own perceptions and actions in the process of establishing Knowledge Management. This leads to an understanding of Knowledge Management as an autonomous venturing process' (p. 282). The focus of this model is on the process of conducting KM on the actions of the employees at the organizational level. These scholars found that there is a 'process of establishing KM as a member-driven start up process, which emerges and develops outside the scope of the current corporate strategy' (p. 287). They named it 'KM venturing' comprising of the 'creating process' 'process of negotiating' and 'formalizing process' as given in Fig. 1.7.

This model emanated from a situation where the members of the organization themselves started the process of the establishment of Knowledge Management. This model pinpoints that it is different from other process models on four counts.

Fig. 1.7 Process model

'Firstly, it is a process study as opposed to a variance study. This implies focus on how events, organizational members and context interact and unfold overtime rather than on relationship between dependent and independent variables and subsequent results. Secondly, it describes a bottom up, employees driven process as opposed to a top-down management driven process. Thirdly, it puts a strong emphasis on the interrelationship between meaning construction and action rather than focusing on action alone. And fourthly, it shows how an organization's identity and culture have significant implications for the process of establishing KM' (p. 295).

1.6.3.2 Process-Centred Knowledge Model and Enterprise Ontology for the Development of Knowledge Management System

The enterprise knowledge object for a process-centred knowledge model is classified into two types: process knowledge and task support knowledge. Han and Park (2009) proposed their 'process-centred model and enterprise ontology for the development of Knowledge Management System' and they thought that 'all domain concepts are related to the process', concept both directly and indirectly (p. 7442). The scholars in their research concluded that 'in order to survive in today's competitive environment, most enterprises recognize the importance of knowledge assets for attaining enterprise performance goals. However, when knowledge is separated from the context of the business process; it cannot contribute to performance goals, because knowledge is utilized and created during the execution of business processes' (p. 7446).

In order to reach their conclusion, the scholars examined the place of their model in overall value-chain context, the structure of their process-centred knowledge model and process-centred enterprise ontology. These three frameworks have been culled out from their model one by one in the following lines.

'In order to maintain sustainability, it is essential to establish a process-centred enterprise structure in a value chain context' (p. 7441). This has been delineated in Fig. 1.8.

'In an agile and flexible structure process is a core element of business operations. Therefore, business process execution: knowledge feeds performance for higher achievement, and performance leads a process to attain business goals. In turn, a business process measures performance through metrics; performance feeds back knowledge for the later use of that knowledge, and knowledge drives the process to improve day-to-day operations' (p. 7441).

Fig. 1.8 Process-centred perspective

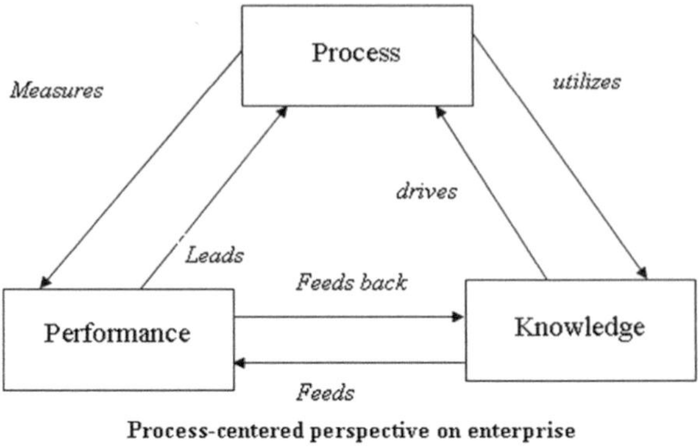

Han and Park (2009) have devised an additional model related to the structure of process-centred knowledge model. They have stated that 'Process Knowledge is acquired and refined during and after process execution, and provides guidelines and tools for task execution' (p. 7442). In this connection, their model as depicted in Fig. 1.9 reifies their ideas.

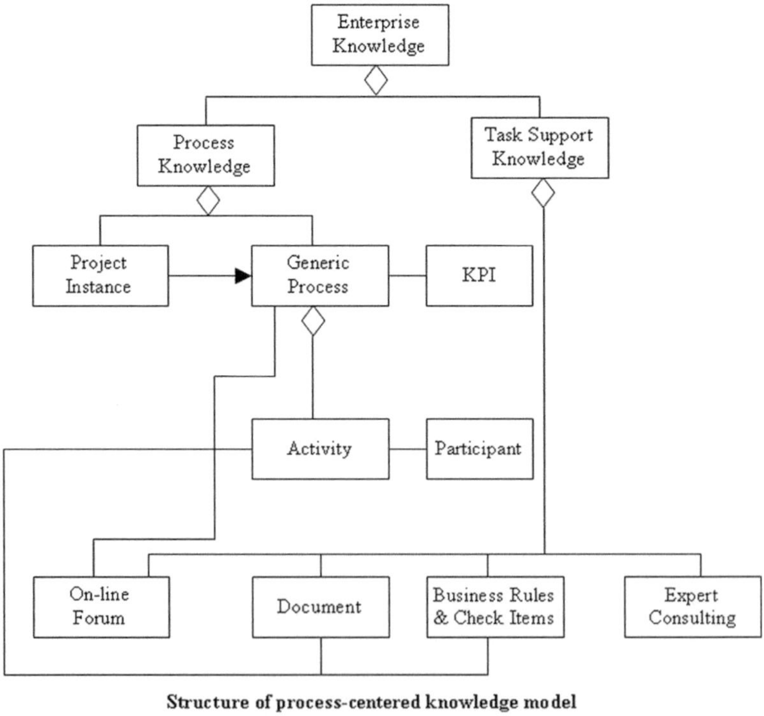

Fig. 1.9 Structure of process

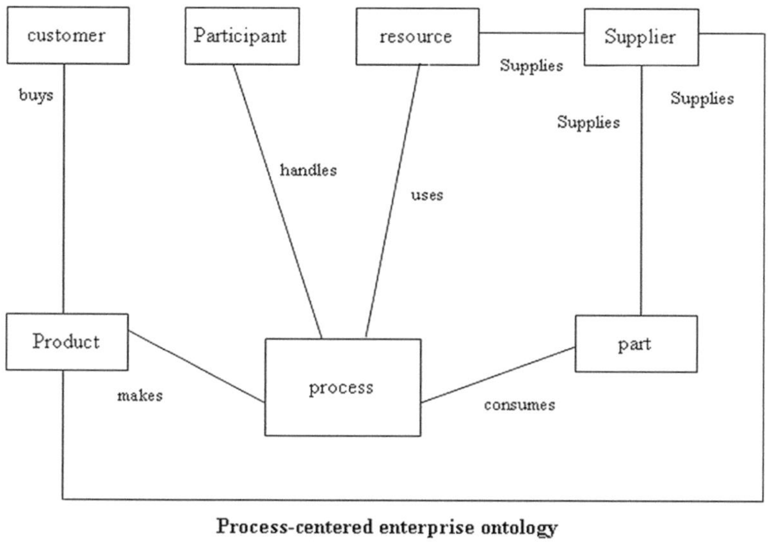

Fig. 1.10 Process-centred ontology

Process knowledge has a KPI (Key Performance Indicator) in order to measure performance and consists of one or more activities, whereas task support knowledge is required in order to make the decisions for target performance during process execution. The scholars lastly proposed process-centred enterprise ontology as delineated in Fig. 1.10.

Figure 1.10 'shows the proposed process-centred enterprise ontology, which represents major enterprise concepts and the relationships between them in a process-centred way. Since the 'process' concept is the element of the process-centred enterprise structure, all domain concepts are related to the 'process' concept directly or indirectly' (Han and Park 2009, p. 7443).

1.6.4 Technological Models

Information and communication technologies have been enabling the Knowledge Management along its various facets of acquiring, creating, collating, updating, disseminating and capitalizing. There is no wonder that scholars have proposed the technological models for Knowledge Management especially with a view to pin-point their requirement at various stages of Knowledge Management. Two representative technological models follow this discussion.

1.6.4.1 The Knowledge Management Technology Stage Model

The firms use Knowledge Management technology for various purposes of Knowledge Management such as intranet, repository, warehousing. The ambition level using Knowledge Management Systems can be defined in terms of stages of Knowledge Management technology. Gottschalk (2006) proposed an 'Expert Systems at Stage IV of the Knowledge Management Technology Stage Model: the case of Police Investigations'. He clarified the systems into four stages: (i) stage I which is labelled end-user tool systems or person to technology; (ii) stage II depicts who-knows-what-systems or person-to-person; stage III is labelled what-they-know systems or person-to-information; and stage IV stands for how-they-think systems or person-to-system. The main objective of a Knowledge Management System is to support the creation, transfer and application of knowledge in organizations. Gottschalk investigated the usefulness of Knowledge Management Systems in police investigations along the four stages of Knowledge Management Systems and finally proposed the model as shown in Fig. 1.11 above.

Gottschalk elaborated the four stages by enumerating the information technology or related gadgets. He narrates that 'Stage I is labelled end-user-tool systems or person-to-technology, as information technology provides people with tools that improve personal efficiency. Examples are word processing, spread sheets, and presentation software. Stage II is labelled who knows-what systems or person to person as people use information technology to find other knowledge workers. Examples are yellow-page systems, CVs and intranets. Stage III is labelled

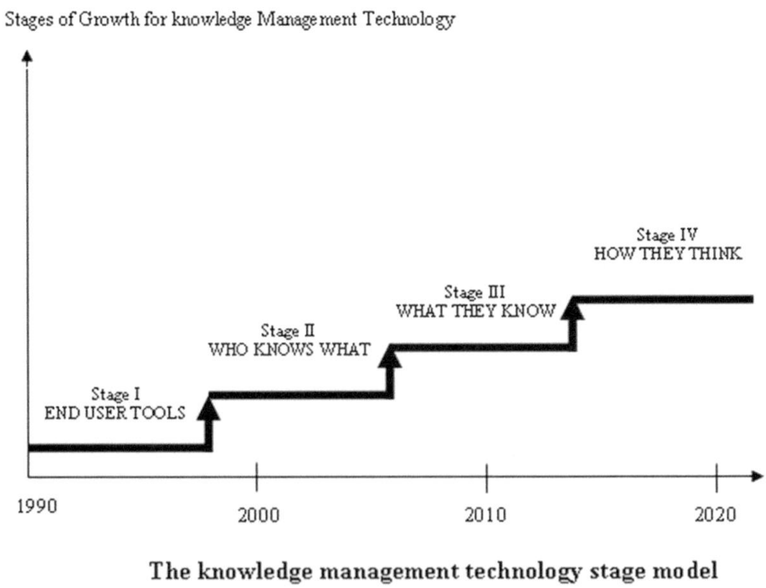

Fig. 1.11 KM technology stage

what-they-know systems or person-to-information, as information technology provides people with access to information that is typically stored in documents. Examples of documents are contracts, articles, drawings, blueprints, photographs, e-mails, presentations and reports. Stage IV is labelled how-they think systems or person-to systems, in which the system is intended to help solve a knowledge problem. Examples are expert systems and business intelligence' (p. 623).

1.6.4.2 Web 2.0 Service Models

The emergence of Web 2.0 concept is, in fact, a network on which peers contribute to the development of tools, content and communities on the Internet. Shang et al. (2011) examined Web 2.0 services that provide different levels of knowledge exploitation and developed a framework for classifying existing service models from a knowledge creation perspective. They termed the 'two types of platforms: experience-socialisation and intelligence-proliferation. These involved four types or service models that we termed as Exchanger, Aggregator, Collaborator and Liberator' (p. 178).

These scholars referred to SECI models and observed, 'Although people can go through the whole learning process of Socialization–Externalisation–Combination–Internalisation (SECI), various Web 2.0 services have been developed to provide ways to support different stages of this process. We adopted a knowledge-creating perspective to define different types of service models by analyzing the activities of web 2.0 applications' (p. 178). The model proposed by them is given in Fig. 1.12.

The scholars elaborated various Web 2.0 applications in Table 1.3 given ahead on page number 31.

The Web 2.0 service models of exchanger, aggregator, collaborator and liberator need elaboration as reported by the scholars.

'(i) Exchanger: This is a platform that enables knowledge socialization and externalisation with a low control mechanism. It is usually a web site that has instant-messaging functions to facilitate exchange of shared experiences via online communication. This kind of service provides a platform such as MSN or Skype that allows users to exchange information via written or voice messages' (Shang et al. 2011, p. 181).
'(ii) Aggregator: This is a platform that enables the knowledge creating cycle from socialization and externalization to combine with low control mechanisms. Its Website can aggregate syndicated web content, such as news headlines, blogs, podcasts and video logs into a single location for easy viewing. It provides a storage platform such as Facebook, YouTube, or Twitter for sharing user's information in an allocated space that is easily accessible over the Internet' (Shang et al. 2011, pp. 181–182).
'(iii) Collaborator: This is a platform that enables the knowledge creating cycle from socialization, externalization through combination to internalization with high control mechanisms in place. On such a platform, the mode of internalized knowledge is reflected in two forms: recreated contents or recreated applications.

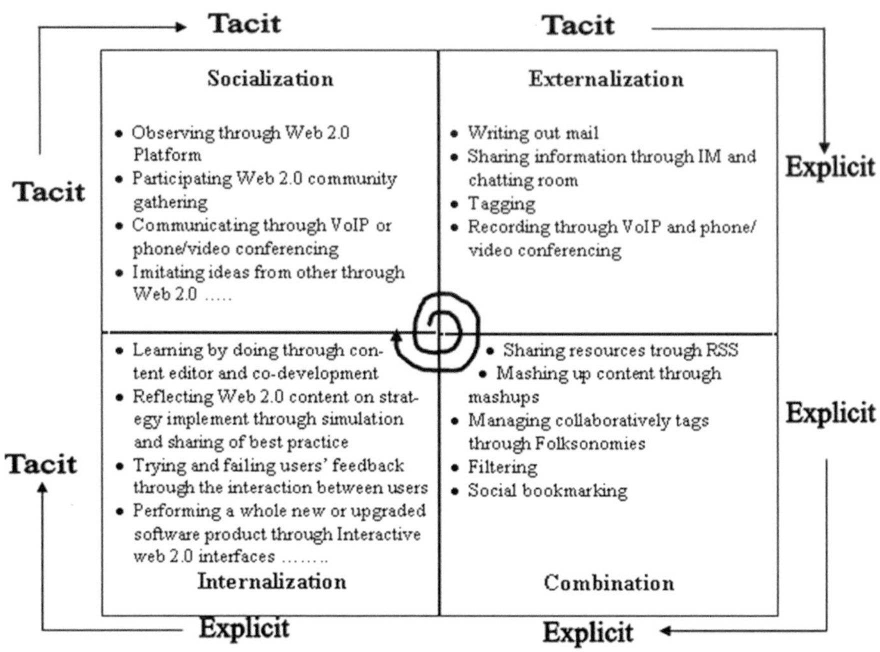

Fig. 1.12 KM creating cycle

The former provides functions for organizing complex information into specific categories and enables participants to review, edit, recreate, and generate contents. Answers.com, salesforce.com, and Yahoo Widget are three examples which offer an API that allows its users to develop applications. These services enable a continuous knowledge-creating process from socializing and externalising knowledge with peers through combining specific knowledge within a collective intelligence to reflecting the knowledge internalization through digitised content or applications' (Shang et al. 2011, p. 182).

'(iv) Liberalisation: This is a platform that provides for the knowledge creating cycles from socialization, externalization through combination to internalisation with low control mechanisms. These communities focus on opening their source code to scrutiny to and to allow upgrading of its quality. Taking Open Office and Linux as examples, users can share the applications they download, as well as revise and update them on the open source community's Web Sites' (Shang et al. 2011, p. 182).

The scholars finally concluded that, 'with the emergence of the Web 2.0 concept knowledge creating services are becoming more complicated and require different efforts from both customers and service providers. Although the application of Web 2.0 is in its early stage, with various applications still progressing, the knowledge creating processes of the platform requires dynamic and contextual management for appropriate knowledge exploitation and to decrease the negative impact of the created knowledge' (Shang et al. 2011, p. 184).

Table 1.3 Web applications, various Web 2.0 applications

Web 2.0 total sites	
Chat (124)	Chat Websites establish Web-based communication among Websites users
E-Mail (59)	Email Websites allow users to manage their email accounts from any computer and offer a Web mail service with a number of features such as connection with photographs, tags, email and voice
Bookmarking (65)	Bookmarking Websites collect, organize and share various types of content that users discover while browsing on the Web
Games (38)	Game Websites provide a browser-based interactive game and allow players to compete against one another and share achievements and scores
Wikis (13)	Wiki Websites allow users to create and share content. Web pages, images and much more in an instant through their browsers
Travel (30)	Travel Websites enable everyone to explore. Clients may want to share advice about local restaurants, or perhaps search the widest possible choice of airline flights and prices
Video (156)	Video Websites allow user to receive Internet Videos
Music (57)	Music Websites are media-centric social networks in which users interact with one other by watching, posting and sharing content of all media types, such as blogs, photographs, audio and video
Programming (38)	Programming Websites enable individuals to generate and share Web 2.0 content code, or other Web applications with one another
Office (43)	Office Websites bring to the Web many of the powerful functionalities of office tools such as MS Word
Blogs (91)	Blog Websites are social networks built around the telling of life stories
Calls and VOIP (17)	Calls and VOIP Websites allow users to call other people using IP software such as Skype
RSS (68)	RSS (Really Simple Syndication) Websites are Web-based, news-feed aggregators designed to allow users to read news from many sources
Photo (99)	Photograph Websites provide online photograph services allowing users to manage and share photographs
Network (115)	A network Website is a social utility for connecting people with friends and others who work study or live around them
Business (18)	Business Websites can help an organization manage and share tasks—e.g. Web-based project management or financial management—in a team or group in an effective and transparent manner using management tools
Web 2.0 tools (11)	Web 2.0 tools Websites offer management tools (e.g. a unique tool to access and manage several DBs through a simple Web interface) through Web 2.0 sites

1.6.5 Other KM Models

An attempt has been made to delineate the typical, representative and popular Knowledge Management models from 1.6.1 to 1.6.4.2, a lot many other models which had been proposed by various scholars for various domains, industries, functions and disciplines could not be mentioned. The number and main propositions of such models besides other details are given in Table 1.4.

Table 1.4 Other KM models

S. No.	Scholars and year	Name of the model	Salient features
1	Dufour and Steane (2007)	Implementing Knowledge Management: a more robust model	This model reviews the dominant approaches to KM implementation. 'It looks at the past academic and business practice literature on Knowledge Management and calls for a radical change in the way of thinking and studying Knowledge Management Implementation, which incorporates configuration and contextual theories' (p. 68)
2	Hsieh et al. (2009)	The construction and application of Knowledge Navigator Model (KNM): an evaluation of Knowledge Management maturity	The Knowledge Navigator Model (KNM) develops to navigate Knowledge Management Implementation journey. 'The KNM comprises two frameworks: evaluation and calculation framework' (p. 4087). This model contributes to the Knowledge Management research field by providing five distinct stages of Knowledge implementation, viz., knowledge chaotic stage, knowledge conscientious, formal implementation of Knowledge Management Initiative, Knowledge Management advanced stage and Knowledge Management integration stage
3	Othman et al. (2011)	A novel statistical model assessing the self-performance of Knowledge Management within SMEs in China	This model introduces a new evaluation model by partitioning the process of implementing Knowledge Management into 'three stages, (1) including the external and internal environment analysis; (2) Knowledge Management activity planning; and, (3) the Knowledge Management decision making' (p. 1758). 'In the study, the empirically proved model can be used as a comprehensive metric for enterprises to assess or predict their KMS self's performance' (p. 1762)
4	Eftekharzade and Mohammadi (2011)	The presentation of a suitable model for creating Knowledge Management in Educational Institutes (Higher Education)	It is a descriptive model which investigates the role of 'organizational culture, information technology, organizational structure and human resources in order to establish a suitable model for Knowledge Management' (p. 1001). This model also depicts the stages and steps for the implementation of Knowledge Management in higher education
5	Oztemel et al. (2011)	Enterprise Knowledge Management Model (EKMM)	'EKMM provides methods and methodologies to sustain better Knowledge Management within the enterprises. Enterprise level objectives and strategies are dominant and guides the knowledge related activities starting from knowledge planning and establishing suitable infrastructure. The proposed model enforces the practitioners to implement specific knowledge representation schemes in order to create well defined and understood knowledge utilization, implementation and sharing environment. One of the mostly important elements of this model is to handle knowledge and all related activities at the organizational level and propose a highly knowledge motivated culture' (p. 878)

(continued)

Table 1.4 (continued)

S. No.	Scholars and year	Name of the model	Salient features
6	Handzic (2001)	Integrated socio—technical Knowledge Management model: an empirical evaluation	This conceptual model has been proposed with three inter-related concepts (knowledge stocks, processes and socio-technical Enablers). 'The findings provided a confirmatory text of the proposed model and revealed social factors to be of greater importance than technical factors in advancing organizational knowledge in the case of public administration organizations. They also pointed to leadership as the single most important enabler of organization KM in these organizations' (p. 198)
7	Jeon et al. (2011)	An integrative model for knowledge sharing in communities of practice	'This study has identified confirmatory relations by integrating some theoretical models on human behavior designed to clarify the mechanism of knowledge sharing in CoPs. With an understanding of the entire knowledge sharing process (attitude-intention-behavior), companies should establish a spontaneous knowledge sharing culture' (p. 264). This model has attempted to identify the factors and relationships that influence Community of Practice (CoP) member's knowledge sharing attitudes, intentions and behaviours. Whereas both extrinsic and intrinsic motivational factors positively influenced attitude towards knowledge sharing behaviours, intrinsic motivational factors were more influential in this regard
8	Edgar Serna (2012)	Edgar Serna M. maturity model of Knowledge Management in the interpretativist perspective	This model entails that knowledge does not exist independently of human experience, social practice, of knowledge itself and its use, where it is shared by the social practices of communities because it is dynamic and active. 'The interpretativist perspective seems to be more complex to implement and to model by ICTs. However, the interest for taking it is to share the object named knowledge based on the support, mapping, storage understanding and dissemination, to support and to create the many possible activities performed by people through the application of their knowledge. The technological action based on this proposal ranges from the idea that technology can help to "manage knowledge" to the idea that the technologies and approaches for Knowledge Management can improve the different and complex activities of persons dedicated to create knowledge' (p. 6)

1.7 Processes Involved in Knowledge Management

As an organization or for that matter a society proceeds towards knowledge economy, they make their 'living from knowledge creation and knowledge application' (Gottschalk 2007). This movement entails the usage of knowledge in various forms, stages, situations and purposes. If we treat Knowledge Management as the chief driver of knowledge economy ***tabula rasa***, we have to chart out various processes of Knowledge Management. With this in view, this section has been incorporated. The perusal of reading material, articles, books and research reports of various scholars including electronically available literature the Knowledge Management Processes may be delineated under eight heads, viz., knowledge creation, knowledge acquisition, knowledge collation, knowledge storage and use, knowledge dissemination, knowledge sharing, knowledge reuse and synthesis and knowledge capitalization.

1.7.1 Knowledge Creation

The business transactions and activities create a huge amount of data and information. In order to create value and remain in competitive advantageous position, the firms have to create knowledge. 'Creating knowledge requires the existence of a person or group of people who come up with new ideas, new concepts, innovative product or process etc.' (Ceptureanu and Ceptureanu 2010, p. 150). 'Thus, knowledge creation involves a people dimension, technology dimension and the processes that link the people and technology (Wickramasinghe 2006, p. 333).

Geisler and Wickramansighe (2009) summarized this aspect when they narrated that 'two conceptual streams of thought have emerged that describe knowledge creation either as a people or technology centric process' (pp. 44–45). They further report that 'Nonaka's *people centric knowledge spiral* is the most widely used framework for knowledge creation through socialization, combination, externalization and internalization'. 'It is almost an axiom that knowledge creation in firms lies at the heart of competitive advantage that "firms learn", "firms know", etc., have become common place expressions in much of the strategy and Knowledge Management literature. However, it is not firms as such that learn, and firms themselves do not process knowledge. So-called firm-knowledge is composed of knowledge sets controlled by individual assets' (Foss and Mahnke 2003, p. 86).

'Spender's *people centric model*' (Newell et al. 2002) corresponds with Nonaka's tacit knowledge which is a part of SECI model proposed by Nonaka. Newell et al. have also quoted about Blacker's *people-centric approach* which depicts that knowledge can exist in several forms (encoded, embedded, embodied, encultured, embrained). By looking at various people-centric models and approach, we may infer that different organizations require different models and no single model can be quoted 'to fit all' kinds of firms.

In addition to people-centric approach, some researchers have concluded that *technology-centric approach* may also be used by the way of mechanistic method which enables knowledge discovery in data bases through data mining tool. Fuller and Wilson (2006 p. 188–189) reported through their research that 'using case classification accuracy as the criteria neural networks have typically outperformed traditional parametric techniques (e.g. discriminant analysis, logistic regression) as well as other non-parametric approaches (e.g. various inductive learning systems such as ID3, C4.5, CART)'. This also takes care of creation of knowledge by using IF-THEN rules from trained feed-forward neural network used in classifiers. Neural network is basically a form of artificial intelligence in which a computer simulates the way human brain processes information. Possibly, this imposes limitations to neural network technique of knowledge creation and brings to the fore the centrality of the people-centric approach.

The latest research reported in the area of knowledge creation (Desouza and Paquette 2011, p. 99) states that 'the creation of knowledge occurs in many dynamic forms. Most often, it is through humanistic means, such as formal training, living through new experiences, or talking with people who share similar interest. Technical mechanisms also assist in the knowledge creation process, such as Knowledge Management Systems, data warehousing and data mining activities. Therefore, the significance of the employee and supporting Knowledge Management technologies has grown substantially over recent decades. Organisations now live (or die) by their ability to create knowledge, innovate, and generate value with new knowledge'.

1.7.2 Knowledge Acquisition

Knowledge acquisition is an activity which explores the possibility of finding expertise which should be acquired from outside through 'relationship with customers, suppliers, competitors and partners in co-operative ventures' (Probst et al. 2002). It means it is the prerequisite of knowledge acquisition 'to identify the required knowledge domains in alignment with the Knowledge Management strategy, locate the source of this knowledge and acquire the required knowledge' (Shukla and Srinivasan 2002, p. 41). It underlines the importance of the outside view which 'is really a fresh view, which brings with it confidence that new order and sound logic can be applied through knowledge so that an organisation can actually improve. Wherever the outside view comes from it must be based on knowledge' (Matthews 2000, p. 10). Of Late, firms have been seen increasingly interested in inter-firm relations and want to develop proximity with outside world.

Martinez-Canas et al. (2012, p. 64) have reported that 'the resource based view of the firm posits that intangible resources that are valuable, rare, imperfectly imitable and not substitutable in the long term provide a competitive advantage for organization'. They further state that the 'knowledge based view of the firm also asserts that more innovative firms must leverage their acquisition capabilities to update their knowledge capital constantly to match new environmental conditions'. With this in view firms should have their skills and ability through repeated interaction to 'recognize and evaluate' the pertinent external knowledge.

Business intelligence-related technical tools can help companies in acquiring knowledge from other companies besides browsing the *World Wide Web* and *Google scholar* and *Wikipedia* which have become axiomatic these days. Environmental scanning is also required by companies to devise new goals and strategies. However, apropos Knowledge Management the competitive intelligence becomes more useful. Parker and Nitse (2006, p. 44) have brought to the fore this fact which is more clear when they write that 'competitive intelligence (CI) is a process for gathering usable knowledge about the external business environment and turning it into the intelligence required for tactical or strategic decisions. The two are strongly connected because gathered CI has no long-term value unless an effective Knowledge Management Process is in place to turn the information into something usable'.

1.7.3 Knowledge Collation

Collation stands for comparison and analysis of two or more sources of information. Its connotations cover calibration also because calibration involves comparing some measurement against a given standard. The codification and classification of the knowledge objects, synthesizing them, making them meaningful and relevant for the target group and adopting the knowledge to the 'firm specific' needs form the objectives of collation (Shukla and Srinivasan 2002). Knowledge collation is a must before it is relegated to repositories. In fact, two major categories critical and not-critical should be ideally created before the nuances of knowledge collation are brought into play. The personnel who are responsible for collation or systems of collation should have an 'awareness of organizational challenges' (Martins and Meyer 2012).

Knowledge Management Processes require a structured, 'coordinated system for managing knowledge effectively' (Xu and Wang 2006). This is required in both forms of knowledge—tacit and explicit. 'While explicit knowledge can be identified more easily, and systems and procedures can be developed to deal with it, for example through feedback and training, dealing with tacit knowledge poses a greater challenge' (Goldsmith and Pillai 2006, p. 314). These researchers offer a solution to these problems by suggesting that for collation of knowledge two types of memories—transactive and mechanistic may be used. '*Transactive memory* refers to the set of individual memory systems in combination with their intercommunications; transactive memory exists as a property of a group as group members share their memories through their interactions with each other and with external memory devices. *Mechanistic memory* refers to information accessed from mechanical systems. A part of transactive memory can be conceptualized as overlapping with tacit knowledge. Codifying such transactive memory and systematizing it will enable the creation of interactive procedures that could assess the level of miscalibration of users' (Goldsmith and Pillai 2006, p. 314).

1.7.4 Knowledge Storage and Use

Once knowledge has been created, acquired or collated, it is to be stored in repositories so that individuals, groups and organization can have access to it. 'Empirical studies have shown that while organizations create knowledge and learn, they also forget (i.e. do not remember or lose track of the required knowledge). Thus the storage, organization and retrieval of organizational knowledge, also referred to as organizational memory constitute an important aspect of effective organizational Knowledge Management' (Gottschalk 2007, p. 33).

Knowledge storage may be done in hard copy form and electronically enabled tools both. The documents in hard copy form may include 'memos, reports, presentations, articles and so forth and put it into a repository where it can be easily stored and retrieved' (Davenport and Prusak 2000, p. 146). The documented knowledge is 'to be integrated in the natural IT organizational environment' (Levy 2011, p. 588). Researchers (Allameh et al. 2011, p. 1214) have concluded that 'creating new knowledge is not enough and mechanisms are needed to store acquired knowledge and retrieve it when needed. The concept of organizational memory is a great solution in this regard. Organizational memory includes knowledge residing in various component forms that may include written documentation, structured information stored in electronic databases, codified human knowledge stored in expert systems, documented organization procedures and processes and tacit knowledge acquired by individuals and network of individuals'. 'In this stage, the organization should classify the filtered knowledge and add it to the organizational memory' (Benbya et al. 2004, p. 212).

From storage and retrieval perspective, the repositories may be classified into three main categories—(i) external knowledge, (ii) structured internal knowledge and (iii) informal internal knowledge (Davenport and Prusak 2000). Some of the aspects of these repositories should be supported by 'an infrastructure capable of supporting the creation and maintenance (···) and an environment that enables the cultivation and facilitation of knowledge sharing and organizational learning' (Al-Hawamadeh 2005, p. 22). Documentary repository becomes handy in this regard because it is computer-based application for storing and retrieving documents in an organized way. It is generally equipped with a search engine that uses keyword matching and similar techniques to locate and retrieve documents of potential interest to users. Storing knowledge may also be done through a number of other means such as knowledge portal, learning reviews, knowledge cafes after action reviews, document libraries, knowledge clusters, expert-locator and knowledge bases (Wikis). Storage in the electronic memory provides unlimited storage capacity and the facility of digitalization. In addition to these, the storage of

Table 1.5 KM Websites and blogs

KM Websites
www.apqc.org
http://en.wikipedia.org/wiki/knowledge_management
www.knowledge-management-online.com
www.library.nhs.uk/KnowledgeManagement/
www.brint.com/km/
www.kmworld.com/
http://knowledgemanagement.ittoolbox.com/
www.apo-tokyo.org
www.skyrme.com
www.gurteen.com
KM blogs
http://km-consulting.blogspot.com
www.kmedge.org/
http://apintalisayon.wordpress.com/
http://kmwiki.wikispaces.com/km+bloggers
www.knowledge-management-online.com/KM-Blogs.html
http://knowledgeproductivity.blogspot.com

knowledge is also possible now through Knowledge Management Websites and blogs as depicted in Table 1.5.

The greater aim of knowledge storage is, in fact, its use in organization by personnel who require knowledge and want it at the right time and at the right place. In present time, the application and utilization of knowledge is the sole purpose of Knowledge Management so that 'reinvention of wheel' could be avoided and time and cost could be saved. The knowledge life cycle (Wikipedia, as accessed on 21 July 2012) depicts the 'use' of knowledge prominently as shown in Fig. 1.14, in the total life cycle as shown in Fig. 1.13.

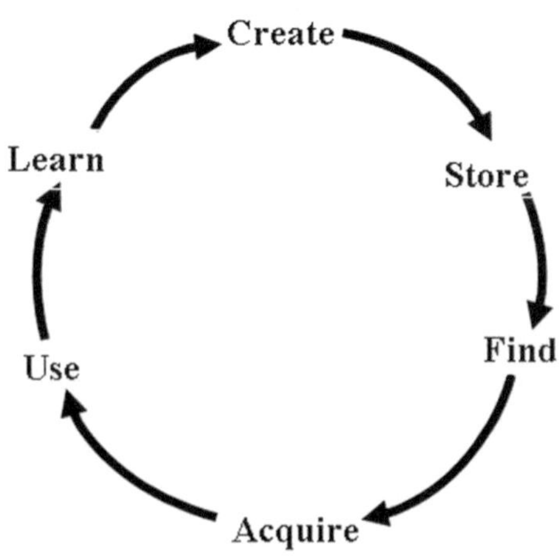

Fig. 1.13 KM life cycle

Fig. 1.14 KM stages

Some of the techniques used by personnel while applying knowledge include peer assist, communities of practice (CoPs), blogs, advanced search, mentor–mentee and apprenticeship. Many of the firms apply 'pull technique' and 'push technique' also for the use of knowledge. 'No matter how many resources an organization pours into creating highly sophisticated knowledge systems, the question looms: will people use them? For any knowledge base or intranet to draw users, it has got to be *better* at helping them find information than the systems and networks with which they are already comfortable and familiar. Having a solid framework and processes for making organizational knowledge available will ultimately ensure that people will use the information' (Bukowitz and Williams 1999, p. 75). O'Dell (2004, pp. 12–13) concluded in her research through a reified figure as reproduced below that, 'Knowledge Management is successful when information and knowledge move through the stages in figure and are actually used'.

1.7.5 Knowledge Dissemination

Knowledge dissemination has been known by various terms such as knowledge translation, knowledge transfer and knowledge exchange. It has been defined as the 'transfer of knowledge within and across settings, with the expectation that the knowledge will be "used" conceptually (as learning, enlightenment, or acquisition of new perspectives or attitude) or instrumentally (in the form of modified or new practices). There are, however, those who see dissemination as having other

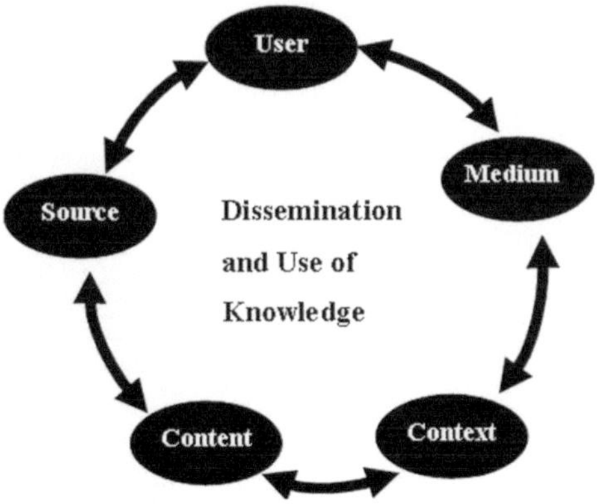

Fig. 1.15 Dissemination & Use of knowledge

legitimate outcomes. Some of these outcomes include: (1) increased awareness; (2) ability to make informed choices among alternatives; and (3) the exchange of information, materials or perspectives' (McKibbon and Lokker 2012). Dissemination is delivering and receiving of knowledge by engaging an individual in the process of transfer of best practices, lessons learned, innovations made or improved processes, etc., within the context of an organization. The relationship of key elements of dissemination leading to knowledge utilization as propagated by Asian Development Bank (2009) is pictorially depicted in Fig. 1.15.

The knowledge dissemination may be done through using Information and Communication Technologies, human-based social interactions and systems or through pathways and media-based means and library.

1.7.5.1 Information and Communication Technology-Based Knowledge Dissemination

Corporate Portal is by far the most popular IT-enabled platform having potential of 'providing organizations with a rich and complex shared workspace' (Benbya et al. 2004). Other scholars (Woods 2006; Hammett and Collins 2002; Tiegland et al. 2000) have also accorded top importance to corporate portal and other IT-enabled sources as the most important tool for knowledge dissemination. Corporate Portals are single-point Web browser interfaces used within organization to promote the gathering, sharing and dissemination of information throughout the enterprise. 'The terms Employee Portals, Enterprise Intranet Portals, Corporate Portals, Business-to-Employee Portals, and Business-to-Employees Systems are sometimes used interchangeably as synonyms to refer to the category of portals, which aim at providing employees with in-time relevant information they need to perform their duties and make efficient business decisions' (Benbya et al. 2004, p. 204).

Groupware and Shareware, virtual conference rooms, bulletin boards and discussion databases are other IT-enabled tools for knowledge dissemination. With the advent of product merged with IT and CT technologies, most popularly known as 'mobile phone' the ICT enabling for knowledge dissemination has reached the zenith. This amalgamation of features of IT and CT has encouraged identifying 'a series of requirements (metrics) from a technology perspective that would enable e-learning for knowledge dissemination in an organization, and how each requirement may influence its adoption by the organisation' (Sivakumar 2006, p. 154).

1.7.5.2 Human-Based Social Interactions and Systems

The end-user where the disseminated knowledge will germinate into 'actionable knowledge' is individual and group. There are two main fora Communities of Practice (CoP) and Community of Interest which help the process of this germination. Communities of Practice are groups of people who share a concern, a set of problems, or a passion about a topic, and who deepen their knowledge and expertise in this area by interacting on an ongoing basis' (Wenger et al. 2002, p. 4) The members of CoPs accumulate knowledge and become informally bound by the value that they find in learning together. As the juxtaposition to CoP, the Community of Interest is basically very informal networks of people who 'communicate, share information and build relationships and reputation'.

People involving activities, such as conference, workshop, seminar, training, colloquium, and andragogy- and pedagogy-related functions, such as teaching, role-play, management games, are powerful disseminators of knowledge and information. The structural arrangements such as cross-functional teams, interdepartmental committees and project teams are also equally helpful in disseminating knowledge.

In addition to the above, pathways including working document, research report, networking, interaction, promotional artefacts, publicizing Policy briefs are also useful in disseminating knowledge. Demonstration is a high-impact disseminator. Even corporate library is still relevant and serves as a repository of knowledge from where knowledge dissemination may take place. Care should be taken that all possible techniques such as 'written, graphical, electronic, print, broadcast and verbal media' are used for dissemination of knowledge. While initiating the process of dissemination of knowledge, the interplay between information users, sources of information, information contents and the medium of information should be planned beforehand to make it a success.

1.7.6 Knowledge Sharing

Knowledge sharing is that process which enables the knowledge held in an individual or groups to be transferred to others in an organization so that it could be applied for the improvements or creation of new products, sources and processes (Van den Hooff and Ridder 2004). The knowledge-sharing process has been categorized into two main types of usage: personal advice and electronic document (Haas and Hansen 2007). The personal advice usage includes the direct contact between individuals when they meet, by phone or via email, whereas electronic document usage covers sharing of electronic documents which can be used as 'stand-alone' resources. However, sharing as a process entails the 'art of placing' knowledge by one at the disposition of others. Its ramifications include that it is a voluntary effort through which one gives and receives knowledge and develops a continuity of previous knowledge to the new knowledge which is to be innovated (Camelo-Ordaz et al. 2011). Knowledge sharing has also been depicted as 'solicited knowledge sharing' and 'voluntary knowledge sharing' (Teng and Song 2011). The environmental antecedents to knowledge sharing such as culture, structures and technology may vary as per the choice of solicited or voluntary types.

It has also been argued that explicit knowledge can be shared easily, whereas tacit knowledge is 'inherently difficult to share because it is socially embedded' and sticky (Osterloch and Frey 2000). In fact, explicit knowledge because of its being already documented can be shared without any problem. However, tacit knowledge requires the help of mentor–mentee, apprenticeship, person-to-person contact, face-to-face communication, deputizing and other factors such as trust, respect and friendship in order to be shared. Two important tools, namely narrative and story, have been found to be of immense help in sharing the tacit knowledge (Snowden 2006; Connell 2006). A narrative is an account of connected events. It may also be said that it is the part of a fictional work that tells the story, as distinct from the dialogue. Snowden (2006, p. 680) has stated that, 'one of the important aspects of narrative and one of the reasons it works is that it creates a higher resource of new concepts of ideas with existing patterns than exist in the human mind'. The narrative explains three aspects of knowledge–information flow, namely abstraction, codification and diffusion. It is because of these properties that narratives become so handy for sharing the tacit knowledge. Narratives have also been found as 'sense-making devices' presenting holistic practices and their context.

In juxtaposition to narrative, 'stories can be seen as one of the ways in which we can encode data about our environment, both personal and organizational. A particular strength of storytelling for Knowledge Management lies in its capacity not only to represent such sets of data, but also to offer some insights into the complex interrelationships between such data elements. In an organizational context, these interrelationships might help us to make sense of the organization' (Connell 2006, p. 721). A distinction has been drawn between 'the notion of narratives and the concept of story by taking "Story" as the more specific and narrative as the basic and more general notion. This means that all stories are narratives but not all narratives are stories' (Geiger and Schreyogg 2012, p. 99).

'Knowledge Translation' is yet another tool which is gaining momentum for sharing of knowledge especially in Canada in the realm of clinical discourse. 'Some define the sharing of experiential knowledge as "Knowledge Management" and the sharing of research facts and findings as knowledge transfer, *translation*, exchange, brokering, interaction or mobilization' (Wikipedia 2012a). 'Knowledge mobilization is putting available knowledge into active service to benefit society. It may be knowledge that has been gathered through systematic study or through experience. Both the research regarding knowledge and experiential wisdom are worth sharing to the benefit of others. It is an obligation and a right to share and to have access to beneficial knowledge' (Wikipedia 2012b).

1.7.7 Enablers and Barriers to Knowledge Sharing

At this juncture, it is pertinent to have a look at the Enablers and barriers to knowledge sharing. A number of scholars have delved in detail in these two areas first, and an attempt has been made to delineate the Enablers. Enabler can be useful on two levels—individual and organizational. At individual level, emotional state poised towards knowledge sharing becomes very important. The pride moves on individual towards eagerness and willingness both to share his or her knowledge. The hubristic pride enables an individual to share out of the wish [···] for the collective interest (Williams and De Steno 2008). 'It implicitly assumes a utopian view of benevolent co-operators who voluntarily give up personal knowledge without appropriate award' (Lam and Lambermont-Ford 2010, p. 52). Motivation plays a pivotal role in knowledge sharing. Hedonic and extrinsic motivators reinforce the propensity for knowledge sharing. Hedonic motivator inculcates individual knowledge sharing, whereas extrinsic can help the group or organization to part with the tacit knowledge.

In the case of organization, the knowledge-sharing culture is required. Training and career progression are to be ensured by an organization to exhort the personnel to share the knowledge. Other human resource management-related factors such as incentive, reward and recognition, job-rotation, job-shadowing, acknowledging the contribution of the individual's normative attitude towards sharing are to be placed properly. Organizations can also take recourse to the formation of cross-functional teams by bringing expert and knowledgeable personnel together for a better understanding of the know-how and skill of others. 'Cross-functional co-operation promotes knowledge sharing, because it is associated with the perception of individuals on the collective use of the shared knowledge in pursuing common interest of the team' (Ghobadi and D'Ambra 2012, p. 286). Contextual factors at the organizational level such as 'trust, management support and learning' are equally required to enable people to share their knowledge (Lilleoere and Hansen 2011). An organization may also have to form 'expert networks' to function as catalyst on the demand side, i.e. individual's knowledge-seeking behaviour of people (King 2006).

The barriers to knowledge sharing emanate from cultural differences and problems related to communication. In a path-breaking study, Lindsey (2006, p. 503) prepared an exhaustive list of barriers to knowledge sharing which is quoted in Table 1.6.

In addition to the above, some barriers such as 'no physical proximity to colleagues', 'unusability of knowledge', 'perception of knowledge as power', 'unawareness about who knows the relevant knowledge' and 'thought to be

Table 1.6 Barriers in knowledge sharing

The capacity for absorbing by receiving unit access to the knowledge ambiguity regarding	*Feeling that the knowledge fits current context: fear of*
1. Ethical situations	1. Becoming redundant
2. Knowledge seekers	2. Distortion or omission of information
3. Peers	3. Exploitation once knowledge is shared
4. Reward	4. Losing confidentiality
5. Supervisor support	5. Losing power once knowledge is shared
6. The knowledge-sharing task	6.Losing resources once knowledge is shared
7. Professional goals	7. Penalty if knowledge is shared
8. Appropriate communication mode	8. Risk
Appropriateness of the availability sharing channel comprising of	*Global constraints including culture and the following factors*
1. Dynamic channels to share knowledge	1. High costs of knowledge search
2. Knowledge-sharing technology	2. Hostile attitude toward knowledge sharing
3. Static channels to share knowledge	3. Improper feedback
4. Time to dedicate to knowledge sharing	4. Inability to understand nonverbal cues
Communication direction apropos	5. Inability to voice relevant knowledge
1. Communication frequency	6. Inappropriate physical appearance
2. Communication of organizational vision	7. Informal social groups or cliques
3. Communicators' lack of credibility	8. Information overload
Compatibility of legacy systems and other systems	9. Internal resistance to knowledge sharing know-it-all attitude
1. Sharing systems	*Knowledge sharing structure match compatibility of*
2. Confidence in the knowledge	1. The organization's style
3. Cultural differences	2. Knowledge-sharing system simplicity
4. Defensiveness for gaining knowledge	*Knowledge workers have a local orientation-lack of*
5. Desire to retain information ownership	1. A knowledge-sharing facility
6. Differences in perceptions of workers	2. Clarity and conciseness
7. Discontinuity in progress toward goals	3. Common ground
8. Effectiveness of sharing channel	4. Contextual clues
9. Effectiveness of the sharing system	5. Interest in the subject matter
10. Efficiency of the sharing system	6. Motivation to participate
11. Either-or thinking	7. Reciprocity

(continued)

Table 1.6 (continued)

Emotional reactions to sharing in the context of	8. Subject-matter knowledge
1. Employees can identify the knowledge	9. Trust
2. Richness of transmission channels	10. Understanding of technical language
Existing resources sufficient to share expected	*Willingness to share*
1. Associations with other sharers	1. Local problem constraints
2. Contribution to the organization	2. Measurement of knowledge transfer
3. Recognition for sharing knowledge	3. Memory loss
4. Rewards for sharing knowledge	*Motivational disposition of source*
5. Obvious link between sharing and the business problems	1. Multiple languages used by knowledge workers
6. Operating environmental factors	2. Resistance to change
7. Org. communicates goal achievements	3. Satisfactory content of the sharing transaction
8. Overly competitive attitude	4. Self-interest
9. Passive listening	5. sender must establish legitimacy
Perceived value of source's knowledge	6. Sense making
1. Perception that knowledge will be of worth	7. Skepticism toward sharing.
2. Personality conflicts	*Specialization of jobs*
3. Physical distance between workers	1. Specialized languages and methodologies
4. Physical noise and distractions	2. State of mind
5. Lack of poor communications skills	3. Status or position
6. Poor organization of ideas	4. Strengthened group identity
7. Poor spatial arrangements	5. Strengthened personal responsibility
8. Power and status relationships	6. Structural barriers in hierarchical organizations
9. Prejudice or bias	7. Tendency of the receiver to evaluate
10. Prematurely jumping to conclusions	8. Time limitations
11. Preoccupation with an ongoing task	9. Too many gatekeepers
Other factors	
1. Prevention of free riders	7. Understanding of the context
2. Professional cultures	8. Unit goals
3. Proprietary knowledge	9. Unit subculture
4. Proprietary thinking	10. Unwillingness to listen
5. Questionable accuracy of information	11. Usability of the knowledge
6. Receiver perceives enhanced efficacy	12. Use of a network that motivates participation
	13. User friendliness of knowledge sharing system

perceived as fool' are also reported which hinder the knowledge sharing (Cabrera and Cabrera 2002; McLaughlin et al. 2008). It is to be pointed out here that if the knowledge-sharing context undergoes a change, the barriers to knowledge sharing may also change.

1.7.8 Knowledge Reuse and Synthesis

Knowledge reuse stands for use of knowledge more than once. It entails that it is such a process by which 'an entity is able to locate and use shared knowledge' (Majchrzak et al. 2004). Oshri (2006, p. 487) defines knowledge reuse as 'processes (which) emphasize the centrality of knowledge within an organization by aligning information systems and communication technologies with human activity and organizational mechanism, such as learning processes and organizational structures'. There are four types of reusers of knowledge, viz., 'shared work producers', 'shared work practitioners', 'expertise-seeking novices' and 'secondary knowledge miners'. Besides this, the process of knowledge includes six stages—(i) approach for reuse; (ii) search for reusable ideas; (iii) scanning the reusable ideas; (iv) evaluating reusable ideas; (v) conducting the in-depth analysis of reusable ideas and selecting one which is most appropriate; and (vi) use the idea (Majchrzak 2004).

'There are three major roles in the knowledge reuse process: *Knowledge producer*—the originator and documentor of knowledge, who records explicit knowledge or makes tacit knowledge explicit, *knowledge intermediary*—who prepares knowledge for reuse by eliciting it, indexing it, summarizing it, sanitizing it, packaging it, and who performs various roles in dissemination and facilitation, and *Knowledge consumer*—the knowledge reuser, who retrieves the knowledge content and applies it in some way' (Markus 2001, p. 61). In addition to these, the knowledge reusers have to apply social mining for discovering patterns and codifying reusable experiences (Akoumianakis 2009).

In order to ensure reuse of knowledge, the same is to be synthesized in repositories of the organization. Different types of knowledge reusers perceive the role of repositories in different ways; therefore, the synthesis becomes equally important apropos knowledge. 'The knowledge synthesis is a social as well as an individual process. Sharing tacit knowledge requires individuals to share their personal beliefs about a situation with others. At that point of sharing, justification becomes public. Each individual is faced with the tremendous challenge of justifying his or her beliefs in front of others and it is this need for justification, explanation persuasion and human connection that makes knowledge synthesis a highly fragile process' (Vat 2006, p. 533). The challenge is how to design the infrastructure to enable spontaneous knowledge capture and transfer so as to turn the scattered diverse knowledge into well-structured knowledge for reuse. Knowledge Management architecture accommodating individual–organizational learning and intellectual property management together may offer a viable solution to this conundrum.

The technique of building a 'knowledge tree' includes all knowledge subjects, explicitly marking the branches to be retained and those to be pruned is said to be an important practice in synthesis of knowledge. Finally, knowledge is to be integrated into routines and business processes of the organization as it makes its reuse more facilitating. Levy (2011 p. 588) has mentioned it succinctly when she says that, 'The knowledge documented is to be integrated in the natural IT organizational environment. Yet, it was found to be very convenient, when such infrastructure was not available, to document the information and knowledge using a WIKI system. The WIKI's structure enables usage of a template, enabling light as well as tight connections between the items documented, and is suitable for integration within other existing intranet sites, as each page is a unique URL. Further more, the structure of WIKI suits the nature of the knowledge documented, as it does not cover all topics, rather the more important ones, forming a net with holes in between'.

1.7.9 Knowledge Capitalization

In the twenty-first century, economic and social development depends increasingly on knowledge rather than labour and capital. 'The most common denominator of the changing economic structure is a shift away from an economy driven and governed by material inputs into the productive process and its organization, toward an economy in which the transformations of productive and distributive processes are increasingly determined by symbolic or knowledge based inputs' (Stehr 2005, pp. 124–125). Capitalization, especially of knowledge, has become even more urgent in the present financial and economic crisis. It stands for taking the 'chances to gain advantage' from existing knowledge of a firm. The point of focus today is knowledge and service production with more emphasis being laid on knowledge capital. This capitalization of knowledge stimulates key stakeholders to transform individual and institutional experience and knowledge into capital which can be used in future.

Machlup (2005, p. 46) has pondered over the issue of knowledge capitalization in terms of stock when he states that 'A fundamental distinction is commonly made between stocks and flows, usually with reference to goods, to capital funds, to money. The distinction also applies to knowledge. At any moment of time, there is a stock of knowledge; during any period of time there is a flow of knowledge'. So far as this stock is concerned, it includes the 'recoded knowledge' (Explicit) and 'knowledge in the mind' (Tacit). 'Knowledge as an economic good exhibits major limitations in terms of radical uncertainty, non-divisibility, non-excludability, non-exhaustibility, non-appropriability and non-rivalry in use. Much economic analysis has explored the implications with respect to tradability of knowledge' (Antonelli and Teubal 2010, p. 98). The capitalization of knowledge is usually analysed by taking recourse to external socio-economic factors.

The process of knowledge capitalization involves the selling or producing economic added value in the knowledge for which it must be completely understandable and reproducible by both inventor and others. Viale (2010, p. 31) narrates the process of capitalization in his seminal work by stating that, 'The generation of economic value can be said to be "direct" when one sells the knowledge for some financial, material or behavioural good. The generation of economic value is considered "indirect" when it allows the production of some material or service goods that are sold on market. The direct mode comprises the sale of personal know-how, such as in the case of a plumber or of a sports instructor. It also comprises the sale of intellectual property as in the case of patents, copyrights or teaching. The indirect mode comprises the ways in which organizational, declarative or procedural knowledge is embodied in goods or services. The economic return in both cases can be financial (e.g. cash), material (e.g. the exchange of consumer goods) or behavior (e.g. the exchange of personal services)'. Venture capital is a concrete example of 'knowledge-driven capitalization of knowledge' because it is a capital which is invested in a business project which involves large element of risk that can be taken care of by knowledgeable experts. It shows that knowledge capitalization is an important and probably the end process of Knowledge Management.

1.8 Human Resource Management and Other Organizational and Social Aspects of Knowledge Management

Even before Peter F Drucker propounded the concept of 'knowledge worker' and 'knowledge as resource' (Drucker 2002), scholars working in human resource management and allied fields declared that 'Knowledge Management activities should result in improving productivity, enhancing the business environment and increasing levels of innovation. These activities also may assist organizations to address human resource management problems on local and global levels, and transform human resources managers into knowledge practitioners or facilitators' (Soliman and Spooner 2000, p. 344). A year later, the 'strategic human resource management'-based researchers started incorporating 'Knowledge-based theories of the firm' highlighting that 'firms exist because they better integrate and apply specialized knowledge than do markets' (Mello 2001). Subsequently, while studying the dynamics of human resource and Knowledge Management, Hafeez and Abdelmeguid (2003, p. 153) noted that in their view 'an efficient human resource or intellectual capital investment strategy demands a good understanding of the dynamics of recruitment and training issues'. In the same vein, Marrewijk and Timmers (2003, p. 174) pointed out that 'The HRM function assigns itself a strategical important position as a sounding board for top management, and facilitator and change agent in the restructuring and transformation process'.

In the year 2003, the research was reported on human resources management in practice with 300 models, techniques and tools wherein seventeen models were underlined for organizational learning and Knowledge Management reiterating the sole responsibility of Knowledge Management to human resource management (Kandula 2003). This lineage of treating Knowledge Management as an integral part of human resource management continued unabated over a span of time (Armstrong 2006). Snell and Bohlander (2007) in their textbook 'Human Resource Management' recommended 'Knowledge-based pay' for personnel as already practiced by some companies. The Knowledge Management as part and parcel of human resource management brought the human capital management into focus (Baron and Armstrong 2007) which necessitated strategic human capital management (Ingham 2007). As a corollary to this, human resource transformation was the natural culmination of human resource management so that Knowledge Management could be possible to the fullest extent (Rothwell et al. 2008). Of late, scholars have been reporting that Company-specific 'Knowledge Management System' has systematized the knowledge of employees so that virtual access to it could be possible from any point of globe (Jones and George 2011). Therefore, the study of human resource management aspects of Knowledge Management is 'increasingly useful for HR Managers to take on board as the number of employees carrying out knowledge work grows' (Taylor 2011, p. 204). In a nutshell, we have to study HR issues such as culture, structure, strategy, change management, leadership, communication, motivation, training and development, learning networks, communities of practice (CoPs) and reward and recognition processes apropos Knowledge Management.

1.8.1 Culture

Culture stands for the art, customs, ideas, social behaviour, intellectual achievement of a nation, people or group but we are confining ourselves to the 'organizational culture' in this study. A strong culture gives an organization its stability; however, when looked at from Knowledge Management point of view it may become a deterrent also. 'Organisational culture refers to a system of shared meaning held by members that distinguishes the organization from other organizations. This system of shared meaning is, on closer examination, a set of key characteristics that the organization values' (Robbins and Judge 2008, p. 551). The latest phenomenon is to see organizations as cultures. It entails that 'culture influences the way we think, what we do, how we work, and what is acceptable in the Company environment' (Wikipedia 2012c). This manifests itself in beliefs, stories and experiences of personnel; goals, norms and history of firm; and symbols, rituals and values of the Company. Furnham and Barrie (1993) pointed out twelve types of culture, namely humanistic-helpful, affiliative, approval culture, conventional, dependent, avoidance culture, oppositional, power culture, competitive, perfectionist, achievement and self-actualization culture. Out of all, these humanistic-helpful, affiliative culture, approval and perfectionist culture are important from Knowledge Management point of view. Culture also exists at the levels of assumptions, values and artifacts.

Scholars have found that firms face certain challenges during their lifetime (Fahey and Prusak 1998) and organizational culture is the most difficult among all the challenges. Janz and Prasaruphanich (2003, p. 353) have delineated this fact very candidly. They write, 'Organisational, culture is believed to be the most significant input to effective KM and organizational learning in that corporate culture determines values, beliefs and work systems that could encourage or impede knowledge creation and sharing'. Out of the determinants of values, belief and work systems 'values are more visible' because these are the tight linkage between social group behaviours. It has also been argued that value orientations such as trust and collaboration will lead to greater willingness among firm members to share insights and expertise with each other. In contrast, value systems that emphasize individual power and competition among firm members will lead to knowledge hoarding behaviours (Alavi and Leidner 2002).

Since culture plays a pivotal role in Knowledge Management, it is required to change it to suit the requirement of sharing the knowledge and other knowledge processes, and dissuade personnal for discounting those cultural values which are not in consonance with it. The volunteering behaviour should also be inculcated in workforce. These can be done either through creating a new organizational culture or by maintaining such a culture if it exists. The change of culture may also be taken recourse to provided it requires breaking from old traditions and practices which are hindering the KM facilitating culture. In this regard, four types of approaches have been suggested to change the culture—aggressive, conciliative, corrosive and indoctrinative (Bate 1995). The first two types do not suit Knowledge Management because they lead to 'crisis of change' and bring back the 'status quo', respectively. Corrosive and indoctrinative types usher in 'high participation' and 'socializing' which are conducive to Knowledge Management. However, 'culture diagnosis' and 'culture audit' should precede any intervention of change in culture. Latest research has suggested that 'The ethical and trusting culture has an overarching concern for all members of the organization. Hence, the ethical and trusting culture dimension is common to all, regardless of the internal/external focus or flexibility/order values or the modes of the knowledge creation and conversion process. Unless the environment of mutual trust, compassion and concern for others persists in an organization, creation, conversion and sharing of knowledge will not take place in organization' (Rai 2011, p. 792).

1.8.2 Structure

Organization scientists have been paying keen attention to the 'link between knowledge and organization structure' (Hendricks 2006) for decades. Structure stands for variables such as size, 'degree of specialization' in the tasks assigned to employees and their jurisdiction and dependence on each other. It also has to take care of 'work flow' (organization design, work process redesign and work place design) (Ulrich and Brockbank 2005). Losey et al. (2005, p. 219) reported that their 'research indicates that organizations flourish and performance abounds in organisations that embody winning practices [⋯] which are fundamentally human in nature, and HR must play a critical role in implementing them'. Apropos this structure becomes central in Knowledge Management by providing proper structure to the organization.

There are three types of structures which have been said to be knowledge-friendly basic structures—'team-based organization, the network structure and the hypertext organization'. The main advantage of a team-based structure is that it can integrate various knowledge nuggets in a cohesive unit. Network structure allows maximum freedom to knowledge workers, whereas hypertext organization facilitates amalgamation of flexibility in a 'traditional function structure'. The 'nonhierarchical, self-organizing organizational structures' have also been said to 'encourage sharing and collaboration across boundaries within the organisation' (Gold et al. 2001, p. 188). If the existing structure does not support the Knowledge Management efforts, it could be changed also. It has been reported that, 'redesigning the Company's departmental structure, co-ordination, span of control, reporting relationships, tasks or decision-making procedures (are) relatively quick and direct way to change an organization' (Dessler 2003, p. 218). The latest research has brought out an entirely new form of structure to deal with knowledge intensity, namely 'knowledge-intensive organizations (KIOS)' by which 'Workers have high cognitive skills; workers use novel knowledge to solve complex problems; worker is held highly accountable for decisions; organization produces and sells knowledge; and organization is accredited by a self regulated body' (Makani and Marche 2012, pp. 259–260).

1.8.3 Strategy

Strategy entails planning for achieving a particular 'long-term aim'. Planning per se is done by 'senior management to define a direction or the future state they wish to achieve' (Desouza and Paquette 2011). This gets manifested mostly in vision and mission statements of companies and they are supposed to be most suitable for their future course of action apropos the requirement of businesses. 'To articulate the strategy-knowledge link, a Company must explicate its strategic intent, identify knowledge required to actually execute that strategic choice, and reveal its strategic knowledge gaps' (Tiwana 2000b, p. 153). In fact, while strategizing for Knowledge Management the continuous assessment of existing knowledge is required for its suitability for future needs which involves strategic planning. Lafley et al. (2012) underlined seven steps for 'applying creativity to a scientifically rigorous process' which involves framing a choice, generating possibilities, specifying conditions, identifying barriers, designing tests, conducting tests and making choice. These steps can be taken in both the realms of formulation and implementation (Paiva et al. 2012). 'As knowledge is taking on an important strategic role, numerous companies are expecting their Knowledge Management to be performed effectively in order to leverage and transform the knowledge into competitive advantages. More importantly, the successful Knowledge Management starts with a proper Knowledge Management strategy' (Wu 2008, p. 834).

1.8.4 Change Management

The Knowledge Management is an exercise that can be seen as change management. Change management ideally should be comprehended as an interactive process that links daily work practices with strategic, directed change programmes and performance goals. When Knowledge Management is closely aligned with 'change process, and processed through collaboration, they are not just end measures—They are part and parcel of knowing and getting success' (Fullan 2011, p. 135). The theory of change propounded by Kurt Lewin as quoted by Jones and George (2011) is known as *force-field theory*. This theory depicts a wide variety of forces which come into existence from the structure, culture and control systems of the operation of an organization which do not allow any change. 'At the same time a wide variety of forces arise from changing task and general environments that push organizations toward change. These two sets of forces are always in opposition in an organisation' (p. 361). Managers, therefore, should increase the force to usher in any change. Knowledge Management is no exception. Scholars have talked about two types of changes—evolutionary and revolutionary. In order to achieve the result as required by an organization, management may adopt either of the two types of changes. Prior to this, the management has to assess the environment, lead the change, link it with the strategy and operation of the firm and relate it to the total set of knowledge, skill and attitude of personnel and ensure the creation of primary conditioning features.

Kotter (2007) reported his research underlining eight steps for ensuring change to happen: (i) establish a sense of urgency; (ii) form a powerful guiding coalition; (iii) create a vision; (iv) communicate the vision; (v) empower others to act on the vision; (vi) plan and create for short-term wins; (vii) consolidate improvements and produce still more change; and (viii) institutionalize new approaches. Even if these steps are not taken, there are forces which imbibe change in organizations in due course of time. These forces are the nature of workforce, technology, economic shocks, competition, social trends and world politics (Robbins and Judge 2008). The change management, especially from the point of view of Knowledge Management, will have to surmount the dilemmas of adaptive or rational strategy development; cultural change or structural change; continuous improvement or radical transformation; empowerment or leadership and command; and economic or social goals (Stace and Dumphy 2001). The Knowledge Management perspective demands that 'improvement to processes, particularly communications and human resources practices, plays a pivotal role, as complementary change across all dimensions depends ultimately on the contribution and commitment of organization members' (Graetz and Smith 2005, p. 311).

1.8.5 Reward, Incentive and Recognition

An employee joins the workforce of an organization in anticipation of compensation. This manifests itself in various forms of rewards, incentives and recognition. In modern times, compensation 'includes both monetary compensations (wages and salaries) and non-monetary compensations (appreciation and benefits like cars, notebooks, mobile phones, flexible hours, homeworking, housing, insurance, sick leave, medical and dental care, vacation, holidays, leisure activities training etc.)' (Šikýř et al. 2008, p. 6). People are rewarded for performing duties and doing jobs which increase the likelihood of the organization's business goals being achieved. Apropos this, 'reward strategy is ultimately a way of thinking that you can apply to any reward issue arising in your organization, to see how you can create value from it' (Brown 2001, p. 44). There should be some guiding principles to 'define the approach an organization takes to dealing with reward' (Armstrong 2008, p. 263). The reward system should aim at rewarding 'individuals for their skills, knowledge and competencies relative to their external market value' and 'it should be individualized to fit the characteristics of individuals' (Mello 2001, p. 517).

The rewards can be categorized in many ways. 'At a broad level, an organizational reward system includes anything an employee values and desires that an employer is able and willing to offer in exchange for employee contributions. More specifically, such compensation includes both financial and non-financial rewards' (Cascio and Nambudiri 2010, p. 403). The extrinsic and intrinsic rewards have also been put forth by scholars as two types of rewards. Singh (2003) has delineated this category in a little detail when he states that, 'extrinsic rewards are taken for granted. The challenge lies in creating motivational intrinsic rewards which can address the higher-order needs of employees' (p. 134). This challenge can be surmounted by HR managers by taking the help of conjoint analysis which draws on 'segmentation theories because it establishes where the 'sweet spots' are for which groups of staff there is the greatest scope for generating improved performance or lower staff turnover through HR interventions' (Taylor 2011, p. 154).

Rewards and recognition are based on quality (Pfeffer and Sutton 2000). The knowledge workers ensure quality delivery. 'Thus, performance management of knowledge workers is quite non-traditional. Similarly, knowledge workers require and expect immediate and frequent rewards in recognition of their good work' (Thite 2004, pp. 39–40). Rewards, incentives and recognition should be aligned by organizations to procure the human capital of knowledge workers. Wu et al. (2011) reported their research where they mentioned that 'the reward of Knowledge Management would influence the performance of knowledge worker positively and if the enterprises could timely encourage staff to share knowledge and relate the contribution of sharing knowledge with performance assessment, then better outcome of knowledge sharing would be obtained for sure' (p. 4390).

1.8.6 Training

Training has been seen as one of the four major, basic, fundamental activities of human resource besides sourcing of manpower, their maintenance and separation. 'Training consists of planned programmes designed to improve performance at the individual, group and/or organizational levels. Improved performance, in turn, implies that there have been measurable changes in knowledge, skills, attitudes and/or social behavior' (Cascio and Nambudiri 2010, p. 275). Some authors (Jones and George 2011; Iles et al. 2001) delineate training as part of human resource development and perceive it as learning and depict that these are inseparable. 'Training and development help to ensure that organizational members have the knowledge and skills needed to perform jobs effectively, take on new responsibilities and adapt to changing conditions' (Jones and George 2011, p. 381)

Scholars have classified the types of training in many ways. Some of them bifurcate it in two categories—classroom training and on-the-job-training. Others differentiate it as pedagogy and executive education where pedagogy stands for classroom teaching. In fact, training in the context of Knowledge Management should be seen as andragogy (adult learning) which 'promotes organizational learning and knowledge based on action, integrating more subjective and relational aspects' (Meireles et al. 2012, p. 442). The types of training have also been trifurcated according to levels—organization, team and individual. So far as the training apropos Knowledge Management is concerned, it has to ensure encouraging the information sharing, countering the resistance to disseminate knowledge and information, recognize individual contribution and 'provide security and status' for those individual contributors. It must ensure updation of knowledge, it use, reuse and capitalization. Training should encourage 'Learning to take place through planned and directed routes within an organizational context, and which 'adds value' to an individual's personal development and to an organizations objectives' (Eldrige and McCourt 2003, pp. 236–237).

Like any tool, training with reference to Knowledge Management is for 'improving competence, in turning what you know into what you do. And unless you have already put in a great deal of work there's almost certainly going to be a gap between the two' (Ulrich et al. 2012 p. 216). This entails that training should certainly result in removing this gap altogether. It pinpoints that training imparted should be evaluated. Stone (2011) has rightly concluded that, 'when an evaluation is carried out, a systematic and credible process must be used to collect and analyze data and report findings, conclusions and recommendations. Among other things, the process must include a systematic and comprehensive approach, a framework to define types of data, tools and templates to make the process efficient and a set of standards of guiding principles to promote uniformity and credibility in the collection, analysis, and reporting of quantitative and qualitative data' (p. 20).

1.9 Knowledge Management Technology

Knowledge Management uses information and communication technology for management of tacit, explicit and implicit knowledge and their creation, acquisition, collation, dissemination, use, reuse and capitalization. Knowledge Management technology also helps hastening the processes involved in Knowledge Management and reduces cost and time taken by Knowledge Management. The Knowledge Management System has to take into consideration these aspects.

1.9.1 Knowledge Management Systems

Scholars have used both terms—Knowledge Management Systems (Gottschalk 2007) and Knowledge Management System (Wong and Aspinwal 2006; Yeh 2005) in literature while discussing about information and communication technology used in Knowledge Management. Since the widespread potential of information technology in Knowledge Management Processes is omnipresent, there is no single information system which can cover all requirements of Knowledge Management in a Company. 'Rather Knowledge Management Systems (KMS) refer to a class of information systems applied to managing organizational knowledge for use at the individual, group and organizational level. These systems are IT applications to support and enhance the organizational processes of knowledge creation, storage and retrieval, transfer and application' (Gottschalk 2007, p. 39). Knowledge Management Systems should be implemented in the light of business focus, nature of products or services, culture, the size of firm and availability of resources. 'Typical approaches use information technology to facilitate the collection, storage, manipulation and sharing of knowledge. Building a Knowledge Management System integrated with databases, search and retrieval engines, collaborative tools, groupware or even with intelligent systems is very common' (Wong and Aspinwall 2006, p. 634). The Knowledge Management System has also been seen as a 'framework of an integration of organizational elements in organizational culture, organizational information technology, infrastructure and the organizations store of individual and collective experiences, learning, insights, values, etc.' (Yeh 2005, p. 36). To conclude, Knowledge Management Systems embrace every aspect of Knowledge Management through the use of information and communication technology (ICT). ICT can be bifurcated in the components of computer, its networks and enabled ubiquitous services and its amalgamated interfaces with telecommunication devices.

1.9.2 Basic Computer System Components

These components include hardware, middleware and software and form the computer system. The hardware of computer system equals the collection of physical elements that comprise monitor, keyboard, computer data storage, hard disk, mouse, printer, CPU (graphic cards, memory, motherboard and chips). 'Middleware is computer software that provides services to software applications beyond those available from the operating system' (Wikipedia 2013a). Software is a set of machine-readable instructions which command the computer's processor to perform a specific operation. The basic computer system components are best suited to store the 'knowledge-nuggets' and information. They are rudimentary tools of Knowledge Management technology.

1.9.3 Types of Computer Networks

The computer networks have resulted in these types of genus-intranet, extranet and Internet. An intranet is the generic term for a collection of private computer networks of a Company which operates over the local area network (LAN) that can keep HTML pages. An intranet is accessible only to the employees of the Company. When an Internet is accessible to some external stakeholders in a limited way, the network of extranet becomes operational by associating itself with 'Internet protocols'. 'Intranets and extranets are communication tools designed to enable easy information sharing within workgroups' (About.com 2013).

An intranet enables collaboration, cohesiveness of culture, knowledge sharing through its 'knowledge portal'. Web-publishing is also undertaken by an organization through its intranet which affords knowledge dissemination and sharing besides allowing 'communication within an organization vertically and horizontally' and through this employees can chat, email and or blog with each other. It may pave the way for knowledge use and reuse. An intranet also serves as 'Intranet Wiki' which facilitates knowledge acquisition for the personnel of an organization. Besides serving as bulletin board, an intranet may also provide 'a very interesting groupware function, i.e. groupwork'.

Apart from intranet and extranet, Internet 'is the world-wide network of computers accessible to anyone who knows his Internet protocol (IP) address—IP address is a unique set of numbers' (Net—What Do They Mean? 2013). Internet happens to be 'Network of Networks' consisting of millions of networks in the domains of private, public, academic and business, both locally and globally connected with a broad array of electronic, wireless and optical network of technologies having infrastructure to support email. From the point of view of Knowledge Management besides maintaining Web, Internet allows end number of emails, file transfer, remote computer control and newsgroups. In fact, Internet enables discussion forums, blogs, and social network sites which help dissemination and sharing of explicit, tacit and implicit knowledge across the globe and gives impetus and force to knowledge economy. Access to Internet is possible almost anywhere by numerous means including mobiles.

1.9.4 Ubiquitous Technology and Its Various Forms

The arrival of Internet has made Knowledge Management technology ubiquitous and has helped Knowledge Management in a big way by decentralizing the knowledge and information through search engines and Web-enabled semantic Web and its various formats and versions which lead to knowledge creation, acquisition, collation, sharing dissemination and use. These can be broadly classified into two categories—social networking tools and machine-readable tools.

Social networking sites such as Blogs, YouTube; services such as Facebook and Twitter, Website of Linked In; and Web application platform of Microsoft Share Point are freely available to the knowledge workers for increasing their social capital and acquiring, using and sharing tacit and explicit knowledge is one category. Second category comprises machine-readable applications of RSS feed, Enterprise 2.0, Mashups, Folksonomy and Semantic Web 2.0 and tacit knowledge sensitive Web 3.0. Nevertheless, the tools enumerated in both categories are prominent components of KM Technology which can be used by the help of Internet and intranet (Wikipedia 2013b).

1.9.5 Associated Technologies for Knowledge Management

Two packages ERP and, of late, cloud computing have been enabling the Knowledge Management in a big way by providing the storage, retrieval, work-flow assistance and warehousing for knowledge and information. Accordingly, the role of KM in ERP implementation could be said to encompass methods for creating, organizing, storing and transferring knowledge related to the core organizational processes. This can be achieved through the implementation of a KM strategy which has been defined as, 'A declaration of how the organization will use KM methods, tools, processes and practices to achieve business objectives by leveraging its content, people and processes and how KM will support the organizations overall strategy' (Knowledge Board 2004). In fact, KM and ERP are synergistic solutions (Huang et al. 2002).

Cloud computing service providers allow agencies to use their warehouses for storing information and knowledge and payment is charged as per the actual use and time taken in this endeavour by their clients. Thus, it helps the clients to reduce the cost of maintaining a huge IT infrastructure. 'This technology is primarily driven by the Internet and requires rapid provisioning, high scalability and virtualized environments. It provides the abstractions for the business and is handled by the actual owners of the infrastructure experts. In this demanding world, the raison d'être to adopt cloud computing over standard IT deployment is flexibility, stability, rapid provisioning, reliability, scalability and green solutions' (Saurabh 2011, p. 2).

1.9.6 Information and Communication Technology

Information and Communication Technology (ICT) is not an 'extended synonym for Information Technology (IT). In reality, it stands for the role of 'unified communications and the integration of telecommunications (telephone lines and wireless signals) computers as well as necessary enterprise software, middleware, storage and audio-visual systems, which enable users to access store, transmit and manipulate information' (Wikipedia 2013c). ICT can be useful in lowering the barriers of 'temporal distance, physical and social distance' by using the tools of communication such as video-conferencing, Skype mobile and other 'hand held devices' for buttressing knowledge sharing and helping Knowledge Management.

1.10 Evaluation of Knowledge Management

Knowledge Management is basically a continuous process and its evaluation should be an ongoing process. However, since evaluation is a 'project' the evaluation should define the scope, design, rationale and purpose of evaluation. Fawcett and Rabinowitz (2013) have defined that 'Every evaluation is essentially a research or discovery project. Your research may be about determining how effective your programme or effort is overall, which parts of it are working well and which need adjusting, or whether some participants respond to certain methods or conditions differently from others' (p. 1). It can be deduced from the above that whether Knowledge Management is causing a particular change in a firm including the behaviour of its personnel or not. If changes are taking place, which are the factors ensuring this change. It may also be taken into consideration as whether intended effects have come into existence or not. Which particular method or style was used in the Knowledge Management? If one wants 'reliable answer', one will have to look into the circumstances as well.

1.10.1 Evaluation Design

When the questions for evaluation have been finalized, the choice of design is to be made. A design may throw the reliability to the fore. It also ensures in pinpointing areas which should further be worked on and the areas which are successful and hence may be replicated. The design also adds credibility to the outcome of the evaluation. Identifying the factors which are unrelated to the evaluation may also be underlined. With the help of design, unintended consequences can be identified and corrective measures may be taken in time. A design helps formation of a corrective plan and organizing a structure for evaluation.

1.10.2 Expected Outcome from Evaluation

'One of the primary purposes of the evaluation is to determine if the program or intervention had the desired effect' (SAMSHA 2013, p. 1). Fundamentally, an evaluation is a systematic collection of information about programme activities, characteristics and outcomes to reduce uncertainty, improve effectiveness and make decisions. The expected outcome of an evaluation should pinpoint what is going to be evaluated and who wants to know what from this. The information needed and its collated format is another outcome of evaluation. At the top of all, it should be delineated as what is going to be done with the evaluation results?

1.10.3 Evaluation Mechanism

Knowledge Management Integrated Service Information Platform (2013, p. 1) depicts that KM Evaluation entails 'Enterprise KM evaluation mechanism (…) through which the current KM status can be evaluated objectively'; from this definition, it is clear that if enterprises have to properly comprehend the essence and notion of KM, they will have to adopt a mechanism to know their KM development status through self-evaluation, 'on-site, assessment and online comparison and analysis'. These efforts are aimed to assist businesses in applying KM-related critical technology to further build up highly value-added power brought by knowledge and establish knowledge-based industries that stress sharing, learning and innovation for the challenges of the twenty-first century. This can be said about every industry. Above all every industry will have to adopt the mechanism of knowledge acquisition, creation, sharing, use and reuse and the evaluation mechanism will have to take them into account.

While proposing their mechanism, Knowledge Management Integrated Service Information Platform (2013, p. 3) propounded a structure comprising 'cultural change aspect: 'IT aspect' and 'Process Operations aspect'. The prominence of these three factors must form the main components of the KM evaluation mechanism. They also highlighted that following sixteen 'key areas' of Knowledge Management as mentioned in Table 1.7 should be taken into consideration as part of mechanism.

When we analyse the above structure and KM areas, we find that in reality the 'Knowledge Portal' has become one portal, or for that matter mechanism, through which most of the components of mechanism are made operational (Lee et al. 2013)

Table 1.7 Sixteen key KM areas

S. No.	Key KM areas
1	KM strategy
2	KM promotion
3	KM evaluation
4	Intellectual capital
5	Knowledge identification and classification
6	Knowledge sharing
7	Knowledge capturing
8	Knowledge acquisition
9	Knowledge application
10	Knowledge creation and innovation
11	Knowledge protection
12	Knowledge learning and training
13	Best practices
14	Community of Practice (CoP)
15	IT infrastructure
16	KM system

especially in the three areas of (i) knowledge access; (ii) knowledge creation; and (iii) knowledge transfer. After their research, they concluded that seventeen sub-dimensions of the three areas are to be evaluated apropos Knowledge Portal.

1.10.4 Evaluation of Knowledge Portal

Knowledge Portals are mostly operated on intranet by companies. It is the repository of knowledge from where employees pull the requisite knowledge as per their need. Knowledge Portal allows knowledge access, its creation and dissemination. Lee, Goh and Chu (2013 pp. 6–8) have concluded the dimensions on which evaluation of knowledge may be done—under the following heads and sub-heads of Knowledge Portal:

A. Knowledge Access:

(i) Access to Portal
(ii) Query made and support from portal
(iii) Portal allows results of queries to be further searched
(iv) Portal allows browsing
(v) Portal allows information to be customized
(vi) Organization may also get information customized by portal
(vii) Portal allows access to all those who reach it
(viii) Information is presented in a rich manner (e.g. Images, video, audio, flash) on portal.

B. Knowledge Creation:

(i) Portal has features to capture user information
(ii) Portal is able to collect feedback from users
(iii) Portal is able to acquire domain specific data from users.

C. Knowledge Transfer:

(i) Collaboration from organization to user is possible on Portal (e.g. 'ask an expert')
(ii) Portal allows users to interact (e.g. 'discussion forums')
(iii) Portal offers real-time collaboration (e.g. chat)
(iv) Portal shares additional resources such as links to external Websites and contributions by other users
(v) Portal guides users through online tutorials, demo, hotline, on-screen help
(vi) Portal delivers news and alerts to users through newsletter event calendar, email alert, RSS feed aggregator.

One important way of evaluating the usefulness of Knowledge Portal is by counting the queries made through portal by the help of 'pull technology'. With this in view, effective information storage and management to ensure that the contents on the Knowledge Portals are timely, credible and accurate is a major responsibility of those who manage day-to-day usability and ability of the portal. A knowledge portal becomes pivotal as it is a storehouse of structured information, contains knowledge networks and communities, and provides collaborative workspaces to better encourage, surface and transfer a more spontaneous exchange of tacit knowledge. In a nutshell, it can safely be said that if the evaluation of Knowledge Portal is done properly, it will suffice the evaluation of Knowledge Management of an organization.

1.11 The Knowledge Audit

Knowledge audit has got many benefits for Knowledge Management. It also acts as 'a tool for evaluating knowledge transfer before it occurs' (Mearns and Toit 2008, p. 164). Knowledge audit has got two facets—one, an examination of the sources of data, information and knowledge, and second, identification of unmet needs (Lusignan et al. 2005). However, it should be noted that knowledge audits are tailor-made and suit only to the organization for which they are meant. In his research article titled 'Ten Steps towards Effective Knowledge Audits', Tong (2005) concluded the way knowledge audit was conducted at cultural villages in South Africa. By taking cues from this article, it may be summarized that knowledge audit pinpoints to 'Questionnaire and interviews', 'to assess the knowledge needed', 'original scope of audit remains in main focus', 'audit is conducted as a part of bigger whole', 'audit will differ according to the organizations specific needs', 'focusing on select parts', 'ensuring the correct approach', 'participatory techniques for learning', 'aligned with the owners of the process' and 'knowledge audits will have value-provided results which can be used'. If these criteria are followed, knowledge audit can function as an appropriate tool for evaluation of Knowledge Management in an organization.

1.12 The Structure of Study

The research flow from start to report writing had various steps. These steps were in the commingling of quantitative and qualitative data collection and interpretation and close interaction with the experts. The remainder of the study has been organized in forms of the context, The Description of Case, Research Methodology followed by Data Collection, Analysis and Interpretation and has been concluded with Discussion and Conclusion.

References

Aaron, J.H.: Epistemology and knowledge management. In: Schwartz, D.G. (ed.) Encyclopedia of Knowledge Management, pp. 167–169. Idea Group Inc. PA, USA (2006)

About.com: http://compnetworking.about.com/od/filetransferprotocol/intranets_and_Extranets.htm (2013). Accessed 04 Jan 2013

Akoumianakis, D.: Practice-oriented toolkits for virtual communities of practice. J. Enterp. Inf. Manage. **22**(3), 317–345 (2009)

Alavi, M., Liendener, D.: Knowledge management and knowledge management systems: conceptual foundations and research issues. MIS Q. **25**(1), 107–136 (2001)

Alavi, M., Liender, D.E.: Knowledge management systems; issues, challenges and benefits. In: Barnes, S. (ed.) Knowledge Management System: Theory and Practice. Thomson Asia Pvt. Ltd., Singapore (2002)

Al-Hawamadeh, S.: Knowledge management: cultivating knowledge professionals. Chandos Publishing, Oxford (2005)

Ali, I., Warne, L., Pascoe, C.: Learning in Organisations. In: Schwartz, D.G. (ed.) Encyclopedia of Knowledge Management, Idea Group Reference, Hershay, London, Melbourne, Singapore (2006)

Allameh, S.M., et al.: Examining the impact of KM enablers on knowledge management processes. In: Procedia Computer Science, vol. 3. www.elsevier.com/locate/procedia (2011). Accessed 21 July 2012

Antonelli, C., Teubal, M.: Venture capitalism as a mechanism for knowledge governance. In: Viale, R., Etzkowitz, H.: (eds.) The Capitalization of Knowledge. Edward Elgar, Cheltenham, U.K. Northampton, M.A., USA (2010)

Armstrong, M.: A Handbook of Human Resource Management Practice. Kogan Page, London and Philadelphia (2006)

Armstrong, M.: Armstrong's Handbook of Strategic Human Resource Management. Kogan Page, London, Philadelphia, P.A. (2008)

Asian Development Bank: Disseminating Knowledge Products (2009). google.com. Accessed 27 July 2012

Asian Productivity Organisation: Knowledge Management: Participants Guide, Asian Productivity Organisation accessed through PROQUEST on 1.5.2012 by using 'taxonomy of knowledge' as the keyword for reaching data online (2012)

Aungus, J.: Knowledge Management: Great Concept… But What Is It? in Information Week. http://www.informationweek.com/673/730olkno.htm (2010). Accessed 21 Oct 2010

Awad, E.M., Ghaziri, H.: Knowledge Management, p. 26. Dorling Kindersley (India) Pvt. Ltd., Delhi (2007)

Baloh, P., Desouza, K.C., Paquette, S.: The concept of knowledge. In: Desouza, K.C., Paquette, S. (eds.) Knowledge Management: An Introduction. Facet Publishing, 7, Ridgmount Street, London (2011)

Baron, A., Armstrong, M.: Human Capital Management. Kogan Page, London and Philadelphia (2007)

Baskerville, R., Dulipovici, A.: The theoretical formulation of knowledge management. Knowl. Manage. Res. Pract. **4**(2) (2006)

Bate, S.: Strategies for Cultural Change. Butterworth-Heinemann, London (1995)

Benbya, H.: Knowledge Management, pp. 13–14. Chandos Publishing, Oxford England (2008)

Benbya, H., et al.: Corporate portal a tool for knowledge management synchronisation. Int. J. Inf. Manage. (24) (2004). www.sciencedirect.com

Brelade, S., Harman, C.: A Practical Guide to Knowledge Management, p. 5. Viva Books Private Limited, New Delhi (2006)

Brown, D.: Reward Stragies: From Intent to Impact. CIPD, London (2001)

Bukowitz, W.R., Williams, R.L.: The Knowledge Management Fieldbook, p. 2. Pearson Education Ltd., Great Britain (1999)

Bures, V.: Methodology of knowledge management implementation. Paper presented in international conference on multimedia and ICT in education. Lisbon, Portugal, 22–24 April 2009

Cabrera, A., Cabrera, E.F.: Knowledge sharing dilemmas. Organ. Stud. **23**, 687–710 (2002)

Camelo-Ordaz, C., et al.: The influence of human resource management on knowledge sharing and innovation in spain: the mediating role of affecting commitment. Int. J. Hum. Resour. Manage. **22**(7), 1442–1463 (2011)

Captureanu, S., Captureanu, E.: Knowledge creation/conversion process. Rev. Int. Comp. Manage. **11**(1), 150 (2010)

Carnap, R.: Empiricism, semantics and ontology. In: Stehr, N., Grundmann (eds.) Knowledge: Critical Concepts, vol. I, p. 268. Rutledge, London and New York (2005)

Cascio, W.F., Nambudiri, R.: Managing Human Resources: Productivity, Quality of Work Life, Profit. Tata McGraw Hil Education Pvt. Ltd., New Delhi (2010)

CEN: CEN CWA 14924 European Guide to Good Practice in Knowledge Management—Part 1 to 5 Brussels (2004)

Chen, A., et al.: Effectiveness of heuristic based approach on the performance of indexing and clustering of high dimensional data. J. Inf. Knowl. Manage. **10**(3) (2011)

Cheng, X.: Culture as a special technology in globalization. In: Cullenberg, S., Pattnaik, P.K. (eds.) Globalization, Culture and the Limits of the Market: Essays in Economics and Philosophy, pp. 215–241. New Delhi, Oxford University Press (2004)

Connell, N.A.D.: Organizational storytelling. In: Schwartz, D.G. (ed.) Encyclopedia of Knowledge Management. Idea Group Reference, Hershey, London (2006)

Conway, C., Sligar, C.: Knowledge Management: Unlocking <K_n_o_w_l_e_d_g_e> Assets solutions from Microsoft. Prentice-Hall of India Pvt. Ltd., New Delhi (2002)

Davenport, T.H., Prusak, L.: Working Knowledge, p. xv. Harvard Business Schiool Press, Boston, Massachusetts, U.S.A (2000)

Debowski, S.: Knowledge Management. Wiley India (P) Ltd., New Delhi (2006)

Desouza, K.C., Paquette, S.: Knowledge management: an introduction, pp. 99. Facet Publishing, London (2011)

Despres, C., Chauvel, D.: How to map knowledge management. In: Marchand, D., Davenport, T., Dickson, T. (eds.) Mastering Information Management, pp 170–176. Prentice-Hall, London (2000)

Dessler, G.: Human Resource Management. Prentice-Hall of India Pvt. Ltd., New Delhi (2003)
Drucker, P.F.: Managing in the Next Society. Butterworth Heinemann, Oxford (2002)
Dufour, Y., Steane, P.: Implementing knowledge management: a more robust model. J. Knowl. Manage. **11**(6), 68–78 (2007)
Durmusoglu, S, Jacobs, M., Nayir, D.Z., Khilji, S., Wang, X.: The quasi-moderating role of organizational culture in the relationship between rewards and knowledge shared and gained. J. Knowl. Manage. **18**(1):19–37 (2014)
Easterby-Smith, M.: Contextual Design: A Customer—Centered Approach to Systems Designs. Morgan Kautmann, Naw York, NY, USA (1997)
Edgar Serna, M.: Maturity model of knowledge management in the interpretative perspective. Int. J. Inf. Manage. (2012). doi:10.1016/j.infomgt.2011.12.001. www.elsevier.com/locate/ijinfomgt. Accessed 16 June 2012
Edgar, A., Sedgwick, P.: Cultural Theory, The Key Concepts. Routledge, Landon (2003)
Eftekharzade, S.F., Mohammadi, B.: The presentation of a suitable model for creating knowledge management in educational institutions (higher education). In: Procedia-Social Behavioural Science, vol. 29, pp. 1001–1011 (2011). www.sciencedirect.com. Accessed 17 June 2012
Ein-Dor, P.: Taxonomies of knowledge. In: Schwartz, D.G. (ed.) Encyclopedia of Knowledge Management. Idea Group Reference, Hershey, PA (2006)
Eisenhardt, K., Galunic, K.C.: Coevolving: at last a way to make synergies to work. In: Harvard Business Review on Organizational Learning. Harward Business School Press. USA (2001)
Eldrige, D., McCourt, W.C.: Learning and training. In: McCourt, W., Eldrige, D. (eds.) Global Human Resource Management: Managing People in Developing and Transitional Countries, pp. 236–237 (2003)
Fahey, L., Prusak, L.: The eleven deadliest sins of knowledge management. Calif. Manag. Rev. **40** (3), 265–276 (1998)
Fawcett, S., Robinowitz, P.: Selecting an Appropriate Design for the Evaluation, p. 1 (2013). http://ctb.ku.edu/en/tablecontents/chapter37_Section4_main.aspx. Accessed 10 April 2013
Foss, N.C., Mahnke, V.: Knowledge management: what can organisational economics contribute. In: Smith, M.E., Lyes, M.A. (ed.) The Blackwell Handbook of Organisational Learning and Knowledge Management, p. 86. Blackwell Publishing, Malden, USA (2003)
Fowler, A.: The Role of A1Based Technology in support of the knowledge Management Value Activity Cycle. J. Strateg. Inf. Syst. 107–128 (2000)
Frame, D.: What is epistemology?: the theory of knowledge-is knowledge possible? http://doug-frame-suite101.com/what is epistemology—a18352 (2010)
Frame, D.: The Theory of Knowledge, p.1. www.google.com. Accessed 14 May 2012
Frunham, A., Barrie, G.: Corporate Assessment: Auditing a Company's Personality. Routledge, New York (1993)
Fullan, M.: Change Leader: Learning to Do What Matters Most. Jossey-Bass, A Wiley Imprint, San Franscisco, CA (2011)
Fuller, S.: An overview of social epistemology. In: Stehr, N., Grundmann, R. (eds.) Knowledge: Critical Concepts, vol. II, pp. 40–41. Routledge, London and New York (2005)
Fuller, C.M., Wilson, R.L.: Extracting knowledge from neural networks. In: Schwartz, D.G. (ed.) Encyclopedia of Knowledge Management, pp. 188–89. Idea Group Reference Hershey, London (2006)
Geiger, D., Schreyogg, G.: Narrative in knowledge sharing: challenging validity. J. Knowl. Manage. **16**(1), 97–113 (2012)
Geisler, E., Whickramasinghe, N.: Principles of Knowledge Management: Theory Practice and Cases, pp. 44–45. M.E. Sharpe, Armonk, New York (2009)
Ghobadi, S., D'Ambra, J.: Knowledge sharing in gross-functional teams. J. Knowl. Manage. **16** (2), 285–301 (2012)
Gold, A.H., Malhotra, A., Segars, A.H.: Knowledge management: an organizational capabilities perspective. J. Manage. Inf. Syst. **18**(1), 185–214 (2001)
Goldsmith, R.E., Pillai, K.G.: Knowledge calibration. In: Schwartz, D.G. (ed.) Encyclopedia of Knowledge Management. Idea Group Reference, Hershey, London (2006)

Gottschalk, P.: Expert systems at stage IV of the knowledge management technology stage model: the case of police investigations. Expert Syst. Appl. **31**:617–628 (2006). www.elsevier.com/locate/e&wa. Accessed 16 June 2012

Gottschalk, P.: Knowledge Management System. Idea Group Publishing, Hershey, London (2007)

Gourlay, S.: The SECI Model of knowledge creation: some empirical shortcomings. google.co.in (2012). Accessed 28 May 2012

Graetz, F., Smith, A.: Organizing forms in change management: the role of structures, processes and boundaries in a longitudinal case analysis. J. Change Manage. **5**(3):311–328 (2005)

Graff, T.R., Jones, T.P.: Introduction to Knowledge Management. Butterworth-Heinemaun, Burlington, MA, USA (2003)

Green, C., Ruheler, K.: Globalisation, borderless world and the tower of babel: metaphors gone away. J. Organ. Change Manage. **8**(4):55–68 (1995)

Grover, V.Y., Davenport, T.H.: General perspectives on knowledge management: fostering & research agenda. J. Manage. Inf. Syst. **18**(1), 5–21 (2001)

Gruber, T.R.: Toward principles for the design of ontologies used for knowledge sharing. Int. J. Human Comput. Stud. **43**(5/6) (1995)

Guo, Z., Sheffield, J.: Critical heuristics: a contribution to addressing the vexed question of so-called knowledge management. Syst. Res. Behav. Sci. **24**(6) (2007)

Haas, M., Hansen, M.: Different knowledge, different benefits: towards a productivity perspective on knowledge sharing in organizations. Strateg. Manage. J. **28**:1133–1153 (2007)

Hafeez, K., Abdelmeguid, H.: Dynamics of human resource and knowledge management. J. Oper. Res. Soc. **54**, 153–164 (2003)

Hammett, R., Collins, A.: Knowledge construction and dissemination in graduate education. Can. J. Educ. **27**(4) (2002)

Han, K.H., Park, J.W.: Process-centered knowledge model and enterprise ontology for the development of knowledge management system. Expert Syst. Appl. **36**:7441–7447 (2009). www.elsevier.com/locate/e&wa. Accessed 15 June 2012

Handzic, M.: Knowledge management: a research framework. In: Proceedings of the 2nd European Conference on Knowledge Management (EC KM 2001) Bled, Slovenia (2001)

Hauschild, S., Licht, T., Stein, W.: Creating a knowledge culture. McKinsey Q. **1**(1), 74–81 (2001)

Hawryszkiewycz, I.: Knowledge Management: Organising Knowledge Based Enterprises, pp. 72–73. Palgrave Macmillan, Hampshire, U.K. (2010)

Heisig, P.: Harmonisation of knowledge management—comparing 160—KM framworks around the globe. J. Knowl. Manage. **13**(4), 4–31 (2009)

Hendriks, P.H.J.: Organizational structure. In: Schwartz, D.G. (ed.) Encyclopedia of Knowledge Management. Idea Group Reference, Hershey, London (2006)

Hernandez, J.Z., Sarrano, J.M.: Reflective knowledge models to support an advanced HCI for decision management. Expert Syst. Appl. **19**:289–304. www.elsevier.com/locate/e&wa Accessed 16 June 2012

Van den Hooff, B., Ridder, J.A.: Knowledge sharing in context: the influence of organizational commitment, communication climate and CMC use on knowledge sharing. J. Knowl. Manag. **8**, 117–130 (2004)

Holsapple, C.W., Joshi, K.D.: The inseparability of modern knowledge management and computer-based technology. J. Knowl. Manag. **9**(1), 42–52 (2001)

Holsapple, C.W., Joshi, K.D.: A formal knowledge management ontology: conduct activities, resources, and influences. J. Am. Soc. Inf. Sci. Technol. **55**(7), 593–612 (2003)

Hsieh, P.J., Lin, B., Lin, C.: The construction and application of knowledge navigator model (KNM): an evaluation of knowledge management maturity. Expert Syst. Appl. **36**:4087–4100 (2009). www.elsevier.com/locate/e&wa. Accessed 17 June 2012

Huang, J.C., et al.: Enterprise resource planning and knowledge management systems: synergistic solutions? In: Harmindrath, G., et al. (eds.) New Perspective on Information Systems Development: Theory, Method and Practice. Bled, Slovenia (2002)

Huber, G.: Organisational learning; the contributing process and the literatures. Organ. Sci. **2**(1), 88–115 (1991)

Iles, P., et al.: HRM & knowledge management: responding to the challenge. Res. Practices Hum. Resour. Manage. **9**(i):3–33 (2001)

Ingham, J.: Strategic Human Capital Management. Butterworth Heinemann, M.A, U.S.A (2007)

Janz, B.D., Prasaruphanich, P.: Understanding the antecedents of effective knowledge management: the importance of knowledge centered culture. Decision Sci. **34**(2), 351–384 (2003)

Jeon, A., Krin, Y.F., Koh, J.: An integrative model for knowledge sharing in communities of practice. J. Knowl. Manage. **15**(2), 251–265 (2011)

Johnsen, M.: Knowledge management. http://www.maria-johnsen.com/Knowledge-management/ (2012). Accessed 25 May 2012 through Google Scholar

Jones, G.R., George, J.M.: Contemporary Management. McGraw-Hill, Irwin (2011)

Kandula, S.R.: Human Resource Management in Practice with 300 Models Techniques and Tools. Prentice-Hall of India Pvt. Ltd., New Delhi (2003)

Khin, E.W.S., Fatt, C.K.: The role of reflective practitioner heuristic enquiry in institutional learning and research. Int. J. Educ. **2**(1), E6 (2010)

King, W.R.: Knowledge sharing. In: Schwartz, D.G. (ed.) Encyclopedia of Knowledge Management. Idea Group Reference, Hershey, London (2006)

Kjaergaard, A., Kautz, K.: A process model of establishing knowledge management: insights from a longitudinal field study. Sci. Direct/omega **36**, 282–297. www.elsevier.com/locate/omega (2008). Accessed 15 June 2012

Knowledge Board: European Guide to Good Practice in Knowledge Management—Part 5: KM Terminology. CEN Workshop Agreement (2004)

Knowledge Management Integrated Service Information Platform: Knowledge management evaluation mechanism, pp. 1–3. http://km.ekm.org.tw/KMPP/Web/APO_Dafault.aspx (2013). Accessed 12 April 2013

Kotter, J.P.: Harvard Bus. Rev. (2007)

Lafley, A.G., et al.: Bringing Science to the Art of Strategy. Harvard Business Review South Asia, September Issue (2012)

Lam, A., Lambermont-Ford, J.: Knowledge sharing in organisational context: a motivation based perspective. J. Knowl. Manage. **14**(1), 51–66 (2010)

Lambe, P.: The unacknowledged parentage of knowledge management. J. Knowl. Manage. **15**(2), 190–192 (2011)

Land, F., et al.: Theoretical and practical aspects of knowledge management. In: Schwartz, D.G. (ed.) Encyclopedia of Knowledge Management, p. 851. Idea Group References. Hershay, London, Melbourne and Singapore (2006)

Lee, C.S., Goh, D.H.L., Chu, A.Y.K.: Evaluation of hospital portal using knowledge management mechanisms. Accessed on 12/04/2013 at website of Namyang Technological University, Singapore by using keyword 'KM evaluation' through google.com, pp. 6–8 (2013)

Levy, M.: Knowledge retention: minimising organizational business loss. J. Knowl. Manage. **15** (4), 582–600 (2011)

Lilleoere, A., Hansen, E.H.: Knowledge sharing enablers and barriers in pharmaceutical research and development. J. Knowl. Manage. **15**(1):53–70 (2011)

Lin, H.F.: Antecedents of the stage-based knowledge management evolution. J. Knowl. Manage. **15**(1) (2011)

Lindsey, K.L.: Knowledge sharing barriers. In: Schwartz, D.G. (ed.) Encyclopedia of Knowledge Management. Idea Group Reference, Hershey, London (2006)

Liu, W., et al.: Representing Heuristic Knowledge in DS Theory in Wikipedia. google.com (2012). Accessed 24 May 2012

Losey, M., Meisigner, S., Ulrich, D.: The Future of Human Resource Management. Society for Human Resource Management, Alexandrina, Virginia, U.S.A., Indian Reprint by Wiley India, New Delhi (2005)

Lusignan, S.D., et al.: The knowledge audit of the managers of primary case organizations: top priority is how to use routinely collected clinical data for quality improvement. Med. Inf. Internet **30**(i):69–80 (2005)

Machlup, F.: Stock and flows of knowledge. In: Grundmanu, R., Stehr, N. (eds.) Knowledge Critical Concepts, vol. III (Knowledge & the Economy). Routledge, London and New York (2005)

Macpherson, P. (2007) Unified Theory of Knowledge. http://archive.org.stream/UnifiedTheoryofKnowledge/UTKv6.2_djvu.txt accessed through google.com on 24.05.2012, 9.13

Majchrzak, A., et al.: Knowledge reuse for innovation. Manage. Sci. **50**(2), 174–188 (2004)

Makani, J., Marche, S.: Classifying organisations by knowledge intensity—necessary next steps. J. Knowl. Manage. **16**(2), 243–266 (2012)

Malone, D.: Knowledge management: a model for organisational learning. Int. J. Account. Inf. Syst. **3**:11–123 (2002) (Elsevier Science Incorporated)

Markus, M.L.: Towards a theory of knowledge reuse: types of knowledge reuse situations and factors in reuse success. J. Manage. Inf. Success **18**(1), 57–93 (2001)

Marrewijk, M., Tinmers, J.: Human capital management: new possibilities in people management. J. Bus. Ethics **44**, 171–184 (2003)

Marsick, V., Watkins, K.: The learning organisation: an integrative vision for HRD. Hum. Resour. Dev. Quart. **5**(4) (1994)

Martines-Cañas, R., et al.: Knowledge acquisition mediation of social capital-firm innovation. J. Knowl. Manage. **16**(1) (2012)

Martins, E.C., Meyer, W.J.: Organizational and behavioural factors that influence knowledge retention. J. Knowl. Manage. **16**(1) (2012)

Matthews, M.L.: Knowledge-Driven Profit Improvement: Implementing Assessment Feedback Using PDK Action Theory. Luice Press, London, New York, St (2000)

McGinnis, T.C., Huang, Z.: Incorporation of knowledge management into erp continuous improvement: a research framework. In: Issues Inf. Syst. **V**(2):614–616. google.co.in (2004). Accesed 28 May 2012

McKibbon, K.A., Lokker, C.: http://whatiskt.wikispace.com/knowledge+dissemination (2012). Accessed 23 Sept 2012

McLaughlin, S., et al.: Barrier impact on organisational learning within complex organisations. J. Knowl. Manage. **12**(2), 107–123 (2008)

Mearns, M.A., Toit, A.S.A.D.: Knowledge audit: tools of the trade transmitted to tools for tradition. Int. J. Inf. Manage. **28**, 161–167 (2008)

Meireles, A., et al.: The second generation of knowledge management: an analysis of the relationship between professional training and knowledge management. In: Proceedings of the Europe Conference on Intellectual Capital, Academic Conference Ltd., p 442 (2012)

Mello, J.A.: Strategic Human Resource Management. Wadsworth, Cengage Learning, New Delhi (2001)

Mruthyunjaya, H.C.: Knowledge Management. PHI Learning Private Ltd., New Delhi (2011)

Mvungi, N., Jay, I.: Knowledge management model for information technology support service. Electron. J. Knowl. Manage. **4**(3):353–366 (2010). www.ejkm.com

Net-What Do They Mean? http://www.hcidata.info/inet.htm (2013). Accessed 04 Jan 2013

Newell, S.M., et al.: Managing Knowledge Work. Palgrave, New York (2002)

Newman, B., Conard, K.: A framework for characterizing knowledge management methods, practices and technologies. In: Proceedings of the 3rd International conference on Practical Aspects of Knowledge Management (PAKM 2000). Basel, Switzerland (2000)

Nissen, M.E.: An Extended Model of Knowledge-Flow. Dyn. Commun. Assoc. Inf. Syst. **8**, 251–266 (2002)

Nonaka, I., Takeuchi, H.: The Knowledge Creating Company. Oxford University Press, New York, Oxford (1995)

Nonaka, I.: The knowledge creating company. In: Harvard Business Review on Knowledge Management, p. 26. Harvard Business School Press, USA (1998)

Nonaka, I., et al.: A Theory of organisational knowledge creation: understanding the dynamics process of creating knowledge. In: Dierkes, M., et al. (eds.) Handbook of Organisational Learning of Knowledge, p. 491 (2001)

O'Dell, C.: The Executive's Role In Knowledge Management. Tata McGraw-Hill Publishing Company Ltd., New Delhi (2004)

Oshri, I.: Knowledge reuse. In: Schwartz, D.G. (ed.) Encyclopedia of Knowledge Management. Idea Group Reference, Hershey, London (2006)

Osterloch, M., Frey, B.: Motivation knowledge transfer, and organisational forms. Organ. Sci. **11**, 538–550 (2000)

Othman, A., et al.: A novel statistical model assessing the self performance of knowledge management within SMEs in China. In: Procedia Engineering, vol. 15, pp. 1758–1763 (2011). www.elsevier/locate/procedia. Accessed 16 June 2012

Oztemel, E., Arslankaya, S., Polat, T.K.: Enterprise knowledge masnagement model (EKMM) in strategic enterprise resource management (SERM). In: Procedia Social and Behavioral Sciences, vol. 24, pp. 870–879. www.sciencedirect.com (2011). Accessed 16 June 2012

Paiva, E.L., et al.: Manufacturing strategy process and organisational knowledge: a cross-country analysis. J. Knowl. Manage. **16**(2) (2012)

Parker, K.R., Nitse, P.S.: Competitive intelligence gathering. In: Schwartz, D.G. (ed.) Encyclopedia of Knowledge Management. Idea Group Reference, Hershey, London (2006)

Pasher, E., Ronen, T.: The Complete Guide to Knowledge Management: A Strategic Plan to Leverage your company's Intellectual Capital, p. 188. John Wiley & Sons Inc., Hoboken, New Jersey (2011)

Pawlowski, J., Bick, M.: The global knowledge management framework; towards a theory for knowledge management in globally distributed settings. Electron. J. Knowl. Manage. **10** (1):92–105 (2012). Accessed through google scholar on 21 April 2012

Pedersen, M.K., Larsen, M.H.: Distributed knowledge management based on product state models—the case decision support in health care administration. Decis. Support Syst. **31**:139–158. www.elsevier.com/locate/d&w (2001). Accessed 09 June 2012

Pfeffer, J., Sutton, R.I.: The Knowing-Doing Gap: How Smart Companies Turn Knowledge into Action. Harvard Business School Press, Boston, Massachusetts (2000)

Plass, J., Salisbury, M.W., March, J.: Integrated instruction as a component of a knowledge management system: a case example. Paper presented at the world conference on www and internet (Web Net 2000) November, San Antonio, Tx, USA (2000)

Prat, N.: A hierarchical model for knowledge management. In: Schwartz, D.G. (ed.) Encyclopedia of Knowledge Management, pp. 211–212, 213–217. Idea Group Reference, Hershey, London, Melbourne, Singapore (2006)

Probst, G., et al.: Managing Knowledge Building Blocks for Success. Wiley, West Sussex, England (2002)

Rai, R.K.: Knowledge management and organisational culture: a theoretical integrative framework. J. Knowl. Manage. **15**(5) (2011)

Raybould, B.: Performance support engineering: an emerging development methodology for enabling organizational learning. Perfor. Improv. Q. **8**(1) (1995)

Richter, F.-J., Banerjee, P.: The knowledge Economy in India. Palgrave Macmillan, Yew York (2003)

Robbins, S.P., Judge, T.A.: Organisational Behaviour. Prentice-Hall of India Pvt. Ltd., New Delhi (2008)

Rothwell, W.J., Prescott, R.B., Taylor, M.W.: Human Resource Transformation: Demonstrating Strategic Leadership in the Face of Future Trends. Davies, Black Publishing, Mountain View, California, U.S.A (2008)

Saaty, T.: The Analytic Hierarchy Process. McGraw Hill, New York (1980)

Salisbury, M.: Creating a process for capturing and leveraging intellectual capital. Performance Improv. Q. **13**(3) (2001)

Salisbury, M.W.: Putting theory into practice to build knowledge management systems. J. Knowl. Manage. **7**(2):131 (2003)

SAMSHA: Selecting an appropriate evaluation design, p. 1. http://www.samhsa.gov/access-resources/selecting-appropriate-evaluation-design (2013). Accessed 10 April 2013

Saurabh, K.: Cloud Computing: Insight into New-Era Infrastructure. Wiley India, New Delhi (2011)

Schwartz, D.G.: Encyclopedia of Knowledge Management. Idea Group References, Hershey, London, Melborne, Singapore (2006)

Senge, P.M.: The Fifth Discipline: The Art and Practice of the Learning Organisation. Doubleday/Currency, New York, NY, USA (1990)

Shang, S.S.C., et al.: Understanding web 2.0 service models: a knowledge-creating perspective. Inf. Manage. **48**:178–184. www.elsevier.com/locate/in (2011). Accessed on 15 June 2012

Shorter Oxford English Dictionary: (2007) Oxford University Press Inc., New York, U.S.A

Shukla, A., Srinivasan, R.: Designing Knowledge Management Architecture: How to Implement Successful Knowledge Management Programme. Response Books, A Division of Sage Publication, New Delhi, Thousand Oaks, London (2002)

Shukla, M.: Competing Through Knowledge; Building a Learning Organisation. Response Books, A Devision of Sage Publication, New Delhi/Thousand Oaks/London (1997)

Šikýř, I.M., Boras, D., Bakić–tomić–Tomić, L.: HRM Practices in Managing Knowledge Workers. Paper presented in 2nd special focus symposium on ICE SKS, information, communication and economic science in the knowledge society, Zadar, 13–14 Nov 2008

Singh, S.: Leadership in high performing organisations. In: Bhargava, S. (ed.) Transformational Leadership: Value Based Management for Indian Organizations. Response Books, New Delhi (2003)

Sivakumar, S.C.: e-Learning of knowledge dissemination. In: Schwartz, D.G. (ed.) Encyclopedia of Knowledge Management. Idea Group Reference, Hershey, London (2006)

Smith, G.F.: Towards a heuristic theory of problem structuring. Manage. Sci. **34**(12) (1988)

Snell, S., Bohlander, G.: Human Resource Management, Cengage Learning, Reprinted in India in 2008, New Delhi (2007)

Snowden, D.: Narrative. In: Schwartz, D.G. (ed.) Encyclopedia of Knowledge Management. Idea Group Reference, Hershey, London (2006)

Snowden, D.: Complex acts of knowing; paradox and descriptive self-awareness. J. Knowl. Manage. **6**(2) (2002)

Soanes, C., Hawker, S. (eds.): Compact Oxford English Dictionary. Oxford University Press, Oxford (2008)

Soliman, F., Spooner, K.: Strategies for Implementing Knowledge Management: Role of Human Resource Management. J. Knowl. Manage. **4**(4), 337–345 (2000)

Staats, B.R., Upton, D.M.: Lean Knowledge work. Harvard Business Review, October 2011 Issue, pp. 86–96 (2011)

Stace, D., Dumphy, D.: Beyond the boundaries: Leading & Recreating the Successful Enterprise. McGraw Hill Book Company, Sydney (2001)

Stehr, N.: The texture of knowledge societies. In: Grundmanu, R. Stehr, N. (eds.) Knowledge Critical Concepts, vol. III (Knowledge & the Economy). Routledge, London and New York (2005)

Stone, R.D.: The Real Value of Training, p. 20. Tata McGraw Hill Education Pvt. Ltd., New Delhi (2011)

Szulanski, G., Cappetta, R.: Stickiness Conceptualising, measuring and predicting organization. In: Smith, M.E., Lytes, M.A. (eds.) The Blackwell Handbook of Organisational Learning and Knowledge Management. Blackwell Publishing, MA, USA (2003)

Taylor, S.: Contemporary Issues in Human Resource Management. CIPD, London (2011)

Teng, J.T.C., Song, S.: An exploratory examination of knowledge sharing behaviours: solicited and voluntary. J. Knowl. Manage. **15**(1), 104–117 (2011)

Thite, M.: Managing People in the New Economy: Targeted HR Practices that Persuade People to Unlock Their Knowledge Power. Response Books, New Delhi (2004)

Tiegland, R., et al.: Knowledge dissemination. In: Global R & D Operations: An Empirical Study of Multinational in the High Technology Electronics Industry, Management International Review, Special Issue, 2000/1 (2000)

Tiwana, A.: The Knowledge Management Toolkit. Prentice-Hall International (U.K.) Limited, London (2000a)

Tiwana, A.: Knowledge Management Toolkit: Practical Techniques for Building a Knowledge Management System. Pearson Education, Upper Saddle Review, NY (2000b)
Tong, T.: Ten steps towards effecting knowledge audits. Knowl. Manage. Rev. **8**(3), 5 (2005)
Ulrich, D., et al.: HR from the Outside In: Six Competencies for the Future Human Resource. Tata McGraw Hill, New Delhi (2012)
Ulrich, D., Brockbank, W.: The HR Value Proposition. Harvard Business School Press, Boston, Massachusetts (2005)
Vat, K.H.: Knowledge synthesis framework. In: Schwartz, D.G. (ed.) Encyclopedia of Knowledge Management. Idea Group Reference, Hershey, London (2006)
Vedpuriswar, A.V.: Knowledge Management from A to Z: Concepts, Principles and Contemporary Practice. Vision Books, New Delhi, Mumbai, Hyderabad (2009)
Viale, R.: Knowledge driven capitalization of knowledge. In: Viale, R., Etzkowitz, H. (eds.) The Capitalization of Knowledge. Edward Elgar, Cheltenham, U.K. Northampton, M.A., USA (2010)
Vyas, K: Knowledge Management and Beyond at PwC-OCG. XLRI, Jamshedpur, India (2000)
Warrier, S.E.: Knowledge Management. Vikas Publishing House Pvt. Ltd., New Delhi (2009)
Wenger, E., et al.: Cultivating Communities of Practice. Harvard Business Press, Boston, Massachusetts (2002)
?Whatis.com: Heuristic. http://whatis.techtayet.com/definition/o1,rid9-gci212246,00.html (2012). Accessed 25 May 2012
Wickramasinghe, N.: Knowledge creation. In: Schwartz, D.G. (ed.) Encyclopedia of Knowledge Management. Idea Group Reference, Hershey, London (2006)
Wikipedia: https://en.wikipedia.org/wiki/Wiki_mobilazation (2012a). Accessed 17 Aug 2012 by using keyword "Knowledge translation"
Wikipedia: http://www.stickids.ca/Learnimg/AbouttheInstitute/Programs/Knowledge-Translation (2012b). Accessed 17 Aug 2012 by using keyword "Knowledge translation"
Wikipedia: http://www.entrepreneur.com/article/204890 (2012c). Accessed 19 Sept 2012 by using keyword "Corporate Culture"
Wikipedia: Google.com (2013a). Accessed 07 April 2013 by using keyword "Middleware"
Wikipedia: Google.com (2013b). Accessed 08 April 2013 by using keyword 'components and version' of Knowledge Management Technology
Wikipedia: Google.com (2013c). Accessed on 09 April 2013 by using keyword 'Information and Communication Technology'
Williams, L.A., Steno, D.: Pride and Perseverance: The Motivational Role of Pride. J. Pers. Soc. Psychol. **94**(6), 1007–1017 (2008)
Wong, K.Y., Aspinwall, : Development of a knowledge management initiative and system: a case study. Expert Syst. Appl. **30**, 633–641 (2006)
Woods, S.: Dissemination in portals. In: Schwartz, D.C. (ed.) Encyclopedia of Knowledge Management, Idea Group Reference, Hershey, London (2006)
Wu, C.C., et al.: Drivers of organisational knowledge management. Afr. J. Bus. Manage. **5** (ii):4388–4402 (2011)
Wu, W.W.: Choosing knowledge management strategies by using a combined ANP and DEMATEL approach. Expert Syst. Appl. **35**, 828–835 (2008)
Xu, D., Wang, H.: Integration of knowledge management and e-learning. In: Schwartz, D.G. (ed.) Encyclopedia of Knowledge Management. Idea Group Reference, Hershey, London (2006)
Yeh, Y.M.C.: The implementation of knowledge management system in taiwan's higher education. J. Coll. Teach. Learn. **2**(9):36 (2005)
Yoo, D.K.: Sense—making view of knowledge management: theoretical development and empirical validation. J. Inf. Knowl. Manage. **10**(1), 157 (2011)
Zhuge, H.: A knowledge flow model for peer to peer team knowledge sharing and management. Expert Syst. Appl. **23**:22–30. www.elsevier.com/locate/e&wa (2002). Accessed 11 June 2012

Chapter 2
The Context

2.1 Review of Literature

Knowledge Management-related research at the global, regional and organizational levels is acute (Mohmed et al. 2008). While the potential contribution of Knowledge Management (KM) to a knowledge economy must be precisely assessed, KM can act as a catalyst to facilitate transition to a knowledge-based economy (Castro et al. 2011; Tseng et al. 2011; Siddique 2012). Affective commitment is of particular relevance to the emerging knowledge economy and to the knowledge firms (Casimir et al. 2012), which entails that the study of Knowledge Management should broadly be done at the level of the Company which is the microlevel where Knowledge Management becomes critical. Another important corollary is that as the economy becomes more based on information and knowledge and upon information and knowledge-based products, the diffusion of technological and economic capability around the globe becomes more and more rapid (Koenig and Jank 2012). It has also been reported that 'many of the leading KM theories and practices recognize the importance of social interaction and communications in knowledge sharing and creation. The argument presented here is that such theories and practices do not go far enough' (Crane 2012, p. 457). Fairclough (2001) argues that knowledge-based economies are discourse-based by default. In the light of this, the discourse at the level of the Company should be looked into in-depth.

The approach for the review of the literature has been with reference to the issue whether knowledge is an entity as proffered by epistemological viewpoint or is it an organizational asset which can be managed. The review entails a holistic view of Knowledge Management Implementation. It has also aimed at delineating the causal relation between factors affecting the Knowledge Management Process and practice with particular emphasis on technological parameters and HR parameters. The review has also been carried out apropos various hypotheses/propositions of Knowledge Management Processes and practices with a view to provide a conceptual framework

K.N. Pandey, *Paradigms of Knowledge Management*, Studies in Systems, Decision and Control 60, DOI 10.1007/978-81-322-2785-4_2

for Knowledge Management Implementation. In order to assess the body of literature on Knowledge Management (KM) and information, search was made on five computerized databases (Proquest, Emerald, JSTOR, Elsevier and EBSCO). In addition to these, Google Scholar and Google had also been explored so far as computerized databases are concerned. The keywords used for this search were knowledge as an entity, knowledge as an asset, Knowledge Management Implementation, theories of Knowledge Management, human resource factors of Knowledge Management, technological parameter of Knowledge Management and conceptual framework of Knowledge Management. There were approximately 209 studies referring to these keywords during 1995–2014. Four journals, namely *Journal of Knowledge Management, Journal of Information and Knowledge Management, Harvard Business Review and Organization Science,* were also referred extensively. In addition to these, *Electronic Journal of Knowledge Management* was also perused. Besides this, other journals were also referred for review. In accordance with the objectives to review conceptual and empirical research which entails the contour of the objectives of review, this chapter was written.

The studies have been carried out in European countries apropos Knowledge Management efforts (Akhavan and Jafari 2006), where it has been reported that the strategic planning, information and communication technology, attention to public and private sectors, common reference model, publications about KM, leadership, change management, attention to human resources, organizational learning, horizontal structure, investment on KM, conferences and seminars about KM, and communities of practice, training, culture and pilot studies had been carried out since 1980. However, such a study is not carried out in India. In the literature, we find that for the first time in 2010, a study had been reported on knowledge initiatives in Indian public and private sector organizations. This study pointed out that the Knowledge Management in public sector 'is still in its infancy and has a long way to go in order to keep pace' (Chawla and Joshi 2010, p. 824) with global and continental developments in this area. Nevertheless, the study of two public sector companies, namely BHEL (Chawla and Joshi 2010) and NTPC (Goel et al. 2010), has been carried out and reported in the literature. However, these studies have not been carried out covering all the aspects of Knowledge Management in totality.

The studies reporting about Indian companies have briefly talked about the KM process, KM leadership, building a collaborative learning environment and culture for KM, infrastructure for KM and lastly developing matrix for measurements for continuous improvement in Indian Public Sector. By keeping the above in view, the present study proposes to examine the major aspects of Knowledge Management in Power Grid Corporation of India Ltd. (POWERGRID). The rationale of the study envisages that 'in the continuous process of creating reality, everyday knowledge, routines and interpretations play an important part' (Bergmann 2004, p. 74). It is also proposed to critically assess the relative contribution of KM initiatives, technology and the Knowledge Management from POWERGRID. This is in light with the study conducted for pointing out major lack of strategies designed for the public sector (Cong and Pandya 2003). It entails the conjecture that Knowledge

Management is more than just: knowledge, culture, people, leaders, organizations or technology; it is the sum of all in the proportions necessary to create a fabric that supports the vision of the organization and cannot be prescribed wholesale.

2.2 Scope of the Review

The review of the literature has been carried out from the viewpoint of epistemologies of knowledge in the Knowledge Management literature. It has been examined whether knowledge is an entity as proffered by epistemological viewpoint or is it an organizational asset which can be managed. This has been done to provide a holistic view of Knowledge Management Implementation to help theory-building effort of Knowledge Management Implementation to emerge. The care has been taken to identify the causal relations between factors affecting the Knowledge Management Processes and practices with particular emphasis on human resources and information technology parameters. While reviewing the literature, an attempt has been made to test various hypotheses/propositions of Knowledge Management Processes and practices with a view to provide a conceptual framework for Knowledge Management Implementation in a Company with many business verticals.

2.3 Epistemologies of Knowledge as Entity

The key assumptions and characteristics of the knowledge, as entity, have got two perspectives (Hislop 2013). The first one is the objectivist perspective, and the second one is the knowledge-based perspective of the firm. Within the objectivist perspective, the entitative character of knowledge is the primary characteristic. Knowledge is regarded as an entity/commodity that people possess, but which can exist independently of people in a codifiable form. From this perspective, knowledge can be codified, made explicit and separated from the person who creates, develops and/or utilizes it. Such knowledge can exist in a number of forms, including documents, diagrams and computer systems, or be embedded in physical artefacts such as machinery or tools (Hislop 2013, pp. 17–18).

The knowledge-based theory of the firm represents the dominant theory which adopts the objectivist perspective on knowledge (Hislop 2013, p. 19). Over time, the theory has been developed and refined partly through theoretical development and partly through empirical testing (Berman et al. 2002; Bogner and Bansal 2007; Cuervo-Cazurra and Un 2010; Haas and Hansen 2007; Sullivan and Marvel 2011; Wang et al. 2009). Finally, it is a perspective that underpins much of Knowledge Management literature (Donate and Guadamilas 2010; King and Marks 2008; Stock et al. 2010; Veolpel et al. 2005; Williams 2011). There are two central tenets to the knowledge-based theory of the firm. First, it assumes that knowledge which is difficult to replicate and copy can be a significant source of competitive advantage for firms. Knowledge that is assumed to be difficult to replicate is firm-specific knowledge, which builds from the links to existing knowledge within an organization, and which is related to firm-specific products, services or processes (Wang et al. 2009). Secondly, it assumes that organization provides a more effective mechanism than markets do for the sharing and integration of knowledge between people. Thus, two of the key focuses of research which utilizes the knowledge-based theory of the firm are on the development of firm-specific knowledge (Nag and Gioia 2012) and the relationship between the development and use of such knowledge for firm's performance (Bogner and Bansal 2007; Hislop 2013, p. 19).

The epistemologies of knowledge in the Knowledge Management literature should be further investigated to pinpoint that knowledge-intensive firms are those entities that depend on professional knowledge or expertise corresponding to a specific technical or functional domain (Casimir et al. 2012). It is also to be pinpointed that knowledge should be regarded as a living entity rather than managed as a static object of predetermined process (Chatti 2012). It has been reported that 'In contemporary Anglo-Saxon countries, "epistemologies" is a philosophical term meaning theory of knowledge' (Browaeys 2004). Frame (2010) has further elaborated this aspect by concluding that '*Epistemology* is synonymous with theories of knowledge. For a claim to be true, the claim must have a foundation. Yet finding a firm foundation can be difficult. Epistemology is sub-divided into *rationalism* and *empiricism* which are considered ways of acquiring knowledge. Yet these two concepts have limitations. Truth is essential for one to have knowledge; how can we know something if what they think as true is, in fact, false? (p. 18352)'. The concept of epistemology earlier had been as one which concerns itself with the analysis of what is meant by the term knowledge itself, and with questions about the limits and scope of knowledge, its reliability, and what constitutes justification for holding knowledge (Edgar and Sedgwick 2003).

In fact, epistemological problems and theories are often interconnected with problems and theories in the philosophy of mind. There is, then, much discussion of the topics in the philosophy of mind that are crucial for epistemology, for instance the phenomenology of perceptions, the nature of belief, the role of imagery in memory and introspection, the variety of mental properties figuring in self-knowledge, the nature of interference and the structure of a person's system of belief (Audi 2003). The debate towards the relationship between epistemology and knowledge continued further. In the year 2005, Stanford Encyclopedia of Philosophy–Epistemology reported that defined narrowly, epistemology is the study of knowledge and justified belief. As the study of knowledge, epistemology is concerned with the following questions: What are the necessary and sufficient conditions of knowledge? What are its sources? What is its structure? And What are its limits? As the study of justified belief, epistemology aims to answer questions such as: How we are to understand the concept of justification? What makes justified belief justified? Is justification internal or external to one's own mind?

Understood more broadly, epistemology is about issues having to do with the creation and dissemination of knowledge in particular areas of enquiry.

The issue related with epistemology was reported as entwined with ontology in the year 2012, when it was reported that however, lots of efforts under the semantic Web initiative nowadays brought to public vast amounts of background knowledge which informs of machine readable RDF/OWL ontologies. Little work has actually been done for representing and managing contextual information bounding the validity and applicability of knowledge (Tamilin et al. 2012). Last years are characterized by a rapid growth of publicly available structured knowledge sets whose development has been strongly stimulated by recent semantic Web initiatives. Among numerous databases available nowadays, the most notable are as follows: Freebase, Wikipedia and DBpedia resources which provide high-quality, structured knowledge on well-known topics, and they are famous entities. Nevertheless, this does not negate the importance of knowledge with reference to epistemology where 'the ability to disambiguate a polysemous entity refers that two orthographically different mentions are the same entity and they are in fact, crucial in updating an entity's KB (Knowledge Bases) record. This task has variously called entity disambiguation, record linkage, or entity linking' (Dredze et al. 2013, p. 1). Earlier Han and Sun (2011) proposed the *entity mention model* for the entity linking task. The main advantage of this model is that it can incorporate multiple types of heterogonous entity knowledge. This model also claimed that it has a statistical foundation, making the entity knowledge extraction approach different from most of previous *ad hoc* approaches. Once the entity disambiguation is carried out through *entity mention model,* it comes to the fore that knowledge as an entity has been encompassing the muddling of issues of ontology (the study of being—essentially studying questions of what kind of entities exist) and issues of epistemology (the studying of knowing that essentially is studying what knowledge is and how it is possible) and has been one of the key confusions in various branches of philosophy (Spencer 2012).

The arguments presented so far have established that the epistemological bearings connote that knowledge is an entity, but this entity status can be accorded to the *explicit knowledge* only. So far as the *tacit knowledge* is concerned, it falls beyond the discussion of epistemological scholars and, thus, may be treated separately with a view to find out whether knowledge (inclusive of explicit and tacit) as a whole is an entity. Nonaka et al. (2012, p. 18) have concluded that 'The knowledge based view of the firm views a firm as a knowledge—creating entity, and the capability to create and utilize such knowledge are the most important source of a firm's sustainable competitive advantage. Knowledge and skills give a firm a competitive advantage because it is through this set of knowledge and skills that a firm is able to innovate new products/processes/services or improve existing ones more efficiently and/or effectively. The raison d'être of a firm is to continuously create knowledge'.

Buehl and Alexander (2005) had carried out research regarding knowledge, epistemological competency beliefs and achievement values which were influencing the students and their attitudes towards knowledge as an entity. The researchers stated that 'Cluster analysis and analysis of variance procedures were used to identify students' domain-specific epistemological belief profiles and to examine difference in students' beliefs, motivation, and task performance. Four hundred eighty-two undergraduates completed measures regarding their beliefs about knowledge, competency beliefs, and achievement values relative to history and mathematics and participated in domain learning tasks. Cluster analysis was used to identify epistemological belief profile groups within the domains of history and mathematics. Students with more sophisticated belief profiles had higher levels of motivation and task performance' (p. 967). These researchers took recourse to the study by Hofer and Pintrich (2002) which stipulated that beliefs about knowledge in general influence the domain-general epistemological beliefs which differentiate beliefs about different academic domains. They finally concluded that 'the power of students' beliefs about knowledge and knowing was situated in the classroom. What is significant about the present study is the finding that students operate from a complex epistemological belief system that also entails specific beliefs that can vary according to domain' (Buehl and Alexander 2005, p. 721). It may be deduced from this research that besides treating knowledge as an entity, epistemologists have indicated towards the importance of varying of knowledge according to the context which includes the *tacit knowledge* as well.

2.3.1 Knowledge as an Organizational Asset

On the philosophical level of constructivism, we find that what is common to all constructivist approaches is that they examine the relationship to reality by dealing with constructive approaches in dealing with it. 'For constructivists epistemology, and empirical research based on it, knowledge and the constructions it contains become the relevant means of access to the objects with which they are concerned' (Flick 2004, p. 89). When we move from philosophical pedestal, we find that 'over the last years, there has been an increasing focus on information architecture to help organizations distinguish and manage information as corporate resource' (Xie et al. 2012, p. 125008-1). Initially, researchers were of the view in which information and knowledge were used interchangeably (Ross and Weil 2006). The trend of bifurcating the information and knowledge was started shortly thereafter by the scholars of Knowledge Management. Scholars started using various nomenclatures such as 'Knowledge is the organizational critical asset' (Chen et al. 2012). Dalkir (2011) had reported that even though knowledge cannot be managed as a conventional asset, it is possible to manage intellectual assets, organizational capabilities and processes with a focus on knowledge development and learning. This brought the concept of intellectual assets into the debate. Pacharapha and Ractam (2012) concluded that 'A Company tries to leverage this valuable resource by transferring the existing knowledge within the Company itself as well as from external sources' (p. 724). Since knowledge can reside in an organization and in individuals who are staff in an organization, transferring knowledge can be between organization, between individuals or between individuals and organizations (Sveiby 2001). The knowledge that is embedded within the people and systems of an organization can result in sustainable competitive advantage for the firm because such knowledge is valuable, rare, inimitable and non-substitutable (Afouni 2007).

On a practical front, the knowledge economy is reality. Strategic value derived from Knowledge Management can be created only through collaborative efforts. The relational dimension of an organization, through effective and emotional influences, is an asset that is unique and cannot be replicated. Sustainable competitive advantages can be attributed not only to the ownership of knowledge, but also the ability to proliferate knowledge assets by encouraging knowledge sharing (Reus and Liu 2004). Beyond this, knowledge has also been reported as an intangible private asset which can seduce an individual to withhold knowledge instead of sharing it with others (Webster et al. 2008). By taking a cue from the same logic, knowledge-sharing behaviour implies a conscious act by an individual who participates in the knowledge exchange (Ipe 2003). When individuals perceive the knowledge they possess as a valuable asset, knowledge sharing becomes a process mediated by individuals' decisions about what knowledge will be shared, when to share and with whom to share (Andrews and Delahaye 2000). The other view as found in the literature depicts that knowledge is a strategic asset. It has been reported that 'how organizations develop, store and transfer knowledge is becoming a strategic asset' (Jones and Mahon 2012, p. 774). The researchers have also depicted that knowledge is a crucial organizational asset. The three aspects of knowledge for any organization involve considerations of how the knowledge is obtained, how it is stored and organized and, more importantly, how that knowledge is accessed and shared in real time. Coyte et al. (2012) undertook research of small and medium enterprises in Australia and concluded that knowledge asset can be used as a strategic resource. They concluded that 'SME activities in this area are asset based, since in general, they have less readymade 'infrastructure' for the measurement, management and development and knowledge and other intangible assets' (p. 792).

Researchers also concluded that knowledge is not only an asset, but it can also be measured by using financial techniques. Kamhawy (2012, p. 809) reported that 'Many studies have addressed issues related to evaluating KM performance in organizations. One classical approach… came from the idea of measuring intellectual capital and/or knowledge assets using accounting and finance techniques'. It corroborates that knowledge is treated like any other resource and it has been perceived as the same by scholars. Chatti (2012) conducted a research in Knowledge Management from the perspective of personal Knowledge Management and reported that 'early KM models in the early 1990s shared common emphasis on a static view of knowledge. The knowledge-as-a-thing-driven KM model focuses on the technology-based, predefined representations of knowledge. This model adopts the view of knowledge as an object that can be captured, stored and reused. Thereby, KM is often perceived as merely a technological solution, consequently a significant amount of attention is placed on implementing platforms and repositories to capture, store, control, manage and reuse structured knowledge' (p. 830).

Furthermore, knowledge has been identified as a sustainable competitive advantage which may not only result from tangible assets and resources, but also from knowledge that can be transferred, aggregated and appropriated and that is difficult to imitate (Earl 2001). Harvey (2012) took the knowledge asset to the realm of explicit knowledge, social construct and tacit knowledge and provided a new dimension to knowledge as an asset. He elaborated these aspects by taking cue from Empson (2001) and Brown and Duguid (2001). Since it is seen as an asset, explicit knowledge can be easily transferred from one actor to another and is often referred to as know-what. As a social construct, tacit knowledge comprises each individual's technical and interpersonal skills as well as groups' synergies. Tacit knowledge is commonly called know-how. Predominant in comparison with explicit knowledge, tacit knowledge specifically forms the background necessary to interpret and develop explicit knowledge. However, one should not conclude that there is a sharp division between tacit and explicit knowledge. It has also been reported by Harvey (2012) that each form of knowledge does work that the other cannot and each form of knowledge can often be used as an aid in acquiring the other. Therefore, although the retrieval of expert tacit knowledge is considered quite difficult, Knowledge Management strategies must take both perspectives into consideration. Studies suggest that retrieving, codifying and transferring tacit knowledge is well worth the effort. Among other things, it provides significant cost saving to organizations through improved know-how and increased innovation. However, in spite of recent advances, the dominant belief is still that knowledge can be codified, captured and manipulated instead of, at its core, comprising of the notion of individuals' interaction in which links or ties are considered the bridges by which knowledge transfer occurs between actors.

The one-on-one mentoring and the storytelling group meetings served as a rare instance of the development of a *Ba* (Nonaka et al. 2006). Both of these activities went beyond the mere codification of explicit knowledge as suggested by the 'knowledge as an asset' perspective (Empson 2001) and the source–recipient model. Paramsothy et al. (2013) further expanded this model by depicting its relevance in the banking industry and propagated 'to learn from the explicit and implicit knowledge it possess and turn it into knowledge assets that can generate income stream is a major factor for sustainable competitive advantage' (p. 1350015-1). Witherspoon et al. (2013) further studied the antecedents of knowledge as an asset and emphasized that the 'increasing recognition of the importance of intangible assets including organizational knowledge, to organizational success further motivates the investigation of KS as an organizational control process' (p. 250). It was further suggested that knowledge is the most strategically significant resource of the firm. These notions of the knowledge may, therefore, be considered as an extension of resource-based view of the firm. Bollinger and Smith (2001) and Goh (2002) suggest that knowledge is a strategic asset. Denford (2013) observed that 'A resource is an asset or input to production which an organization owns controls or has access to on a semi-permanent basis' (p. 176). Having reviewed the literature it can be deduced that on the one hand, 'knowledge' has been treated as an 'entity' by scholars and, on the other, it has also been christened as an asset. At this juncture, it is proposed that it should be examined, in the context of a firm—a live case—as whether knowledge is an asset which can be managed as a resource. Therefore, it may be concluded that *If treated as resource knowledge is an asset which can be managed.*

2.4 Aspects of Knowledge Management Implementation

Knowledge Management has largely been reported as the responsibility of human resource management which is enabled by information and communication technology. However, this should be carried out through a suitable strategy. There are two basic strategies for Knowledge Management, and they are codification and personalization. The *codification strategy* uses existing knowledge to solve problems. This strategy uses procedures and knowledge as many times as possible to receive maximum benefits from it. The codification strategy makes more use of information technology (Bowman 2002). The *personalization strategy* uses solutions that are unique to their organization. This often includes customizing their KM tools and software. The investments made by organizations using the personalization strategy promote better ways of sharing between people. KM requires organizations to incorporate the strategy into the daily tasks of the employees. The key Enablers identified are leadership, strategy, culture, technology and measurement. There are KM strategies identified through research and trial and error. There are several valid strategies. Nonetheless, the high–high fit of KM and IT does not always yield positive organizational outcomes, since enough exceptions indicated that business strategy and knowledge strategy (e.g. Asoh 2004), as well as human resource management strategy (Shih and Chiang 2005), are interdependent. Shih and Chiang (2005) noted that fit between KM strategy, corporate strategy and HRM strategy is significantly associated with enhanced KM effectiveness, in terms of process outcome, learning capability and organizational outcomes.

2.4.1 Stages of Knowledge Management Implementation

Organizations should advance through the same series of stages to increase their chances for a successful KM implementation (Dell et al. 2000). Organizations beginning to use the benefits of KM must first perform several tasks. Organizations should define what they want to carry out with the system. They should document what assets they currently have and how they will leverage them. This is *stage one*. The organization explores various choices and experiments with different technologies and strategies in *stage two*. This stage enables the organization to evaluate the strategy to decide how it fits within the culture. Organizations use many different strategies based on their corporate culture. The organization should perform several studies to discover which Enablers and tools work for them. This is *stage three*. *Stage four* expands the KM strategy and analyses where improvements can occur. The implementation is typically enterprise wide and watched closely to ensure continued success. The final stage, *stage five*, incorporates KM into the everyday formation of the organization. Independent management of the KM infrastructures and the information technology infrastructure must occur (Soo et al. 2002). The departments may have the same head, but separated management of them must occur. This arrangement includes a KM department led by a Knowledge Management Officer (KMO). KMOs and other knowledge professional strategies are becoming more common with the increased importance of KM (Elliott and Jacobson 2002).

2.4.2 Categories of Knowledge Management Strategy Implementation

While reviewing the literature, various approaches were found towards the Knowledge Management strategy implementation perspectives. Dufour and Steane (2007) conducted an exhaustive scanning of the approaches which had been in vogue so far as Knowledge Management strategy implementation is concerned. In their conceptual paper, they concluded following four categories of Knowledge Management strategy implementation:

1. The classical approach,
2. The contingency approach,
3. The political approach and
4. The behavioural approach

Since these four categories have been the hallmark of a robust model for Knowledge Management, it requires a little elaboration, which has been adapted form Dofour and Steane (2007). The *classical approach* exists in both the academic and the business practice KM literature. The classical position is founded on the unitarist view of strategy that formulation and implementation can be controlled from one centre of authority and that they are two sequential phases of knowing and effecting strategy. Successful implementation of KM strategy is through a high rational process. The *contingency approach* to KM implementation postures that leadership has a variety of unequal effective structural forms and organizational processes from which to chose in implementing KM strategies. In this approach, the role of the CEO is that of an architect, designing administrative systems to implement change in the management of information and work practices. The contribution of contingency theory applied to organizations was that the unitarist approach is flawed and that organizations comprise many interest groups. The *political approach* to Knowledge Management Implementation is primarily concerned with the impact of patterns of power and influence on the implementation processes and outcomes. The political approach is distinctive in seeing contestation and bargaining as endemic rather than some exceptional or pathological dimension of organizations. Actors engaged in implementation are not so much exhorted to resolve conflict as they try to manage it, within the overall suite of tactics and strategies. The political approach assumes that a plurality of at least partly conflicting interests exists within organizations. The *behavioural approach* will always be an influence on what can be achieved in KM implementation. The behavioural approach incorporates individual and organizational sources of resistance in implementing KM strategies. There is a predominant reliance upon the use of reason or technology in KM implementation and change. However, the features that matter most are those that affect individual motivation, commitment and, in particular, interpersonal cooperation in implementation. Therefore, it may be propounded that *the attitude and behaviour of personnel help Knowledge Management Implementation positively.*

2.5 Factors of Human Resource Management

Human resource management covers areas such as culture, strategy, structure, training, systems, processes and management support apropos Knowledge Management Implementation in organizations. Chen et al. (2012) studied the 'fit' between Knowledge Management and business performance and concluded that human resource management has to play the most important role for Knowledge Management Implementation. Shih and Chiang (2005) also concluded in their research that human resource management is the most critical factor so far as the Knowledge Management Implementation is concerned. Therefore, the review of the literature was conducted with the view to delineate the HR factors including reward and recognition.

2.5.1 Culture

Culture is the collection of central norms that characterize an organization. A corporate culture is reflected in the attitude and values, management style and problem-solving behaviour of its employees. An organization's values, principles, norms and procedures are its cultural knowledge resource (Holsapple and Joshi 2001). An organization's culture is one of the most important factors in effective KM. As will be discussed below, most researchers believe that it is difficult to change the culture, and yet it is almost a priori requirement of effective KM. Two common aspects of culture that have been extensively researched are culture in the national sense of the word and culture in organization often referred to as *corporate culture*. Several researchers have looked at culture in the national sense of the word to better understand how differences in national culture affect information system (Tan et al. 2003). Other researchers have examined culture with respect to the norms and mores within an organization. This research will focus on the second aspect of organizational culture and its interaction with KM programmes and will examine the importance of corporate culture as all pervasive and superimposing over other human resource factors.

Innovation, support, rules and goals were foundation of a cultural Questionnaire focusing on description and evaluation of organizational culture (van Mujjen et al. 1999). A study of four survey instruments for organizational culture found the following five factors: *satisfaction needs* to include behavioural norms and values, *task-oriented organizational growth* defined as a technocratic approach to organizational development, *people orientation* defined as the human factor in a bureaucratic culture, *task orientation* defined as the resistance to new ideas and *positive social relations* in the work place. This and other research using survey instruments for organizational culture studies have somewhat gained acceptance with many scholars in the discipline. If an organization's culture is not appropriate for a knowledge project, no amount of technology, content or project management skills will make the project successful (Davenport and Klahr 1998). In a study of 71 practitioners at KM presentations, culture was perceived to be the biggest barrier to KM implementation (Mason and Pauleen 2003). In their study, Mason and Pauleen operationalized culture as organizational culture, trust, sharing and communication. A culture that supports knowledge is one that values learning and rates experience, expertise and innovation higher than hierarchy (Davenport and Klahr 1998). Organizational culture can influence the adoption of technology (Huang et al. 2003), while cultural drag can dramatically inhibit organizational change efforts (Robey and Boudreau 1999). A culture aligned with organizational objectives benefits all change projects.

Culture is separate from infrastructure (structural and technological), yet culture can still be influenced by infrastructure, as infrastructure can constrain or promote cultural evolution (Holsapple and Joshi 2001). Information technology induced cultural change can be dangerous to an organization when there is a mismatch between the organization's culture and the proposed system (Doherty and Doig 2003). However, the difficulties in changing an organization's culture suggest the cultural limitations on structure and technology may be higher than the infrastructure's limitations on culture. An appropriate knowledge-oriented culture should show a positive orientation towards knowledge sharing and an innovative nature (Davenport and Prusak 1998). A knowledge-sharing culture should already exist if a KMS is to be effective (Damodaran and Olphert 2000). In one study, the authors operationalized culture based on the organization's encouragement of knowledge sharing between employees (Al-Busaidi and Olfman 2005). Most researches that study the elements of a knowledge culture (Levin and Cross 2004; Renzi 2008; Bock et al. 2005) focus on knowledge processes in general without distinguishing between different knowledge processes. Wang et al. (2011) have made a first attempt to focus knowledge culture studies explicitly on the knowledge creation process. Ferraresi et al. (2012) conducted a research regarding innovativeness and performance brought by Knowledge Management and strategic orientation and concluded that 'Thus effective KM can be understood as a set of processes that are embedded in organizational culture and contribute to strategic orientation and innovativeness. The presence of such processes indicates that KM practices are, to a large extent, integrated to the work routines of many companies and can be viewed as part of their culture' (p. 697). Therefore, it may be summarized that *culture is the prime mover of Knowledge Management Implementation.*

2.5.2 *Strategy*

So far as the strategy for Knowledge Management is concerned, the literature has shown various trends. Chen et al. (2012) studied the 'fit' between the business and Knowledge Management and reported the research regarding investigation of various strategies of the organization which is not sufficient. In addition, it is crucial for the analysis and design of the organization as a whole to achieve organizational benefits. In the practical terms, the basic alignment mechanism is by 'strategy', and it is thought that a match between strategy and organization is the key driver to effectiveness at realizing intended strategies. Therefore, this study focuses on three types of strategies discussed previously that are critical to business in today's knowledge-based organization, that is Knowledge Management (KM) strategy and information technology (IT strategy and human resource management (HRM) strategy (Chen et al. 2012, p. 672). In the literature, it has been found that the firms which use human-oriented (personalization) KM strategies must have reward systems that encourage workers to share knowledge directly with others; instead of providing intensive training within the Company, employees are encouraged to develop social networks, so that tacit knowledge can be shared. That is, when human KM strategy is adopted, only the fit between human KM strategy and reward systems of HR strategy is found to have a significant impact on business performance in terms of growth. One possible explanation may be that the strategy a firm used on knowledge sharing in human KM strategy is mainly by members' face-to-face conversation in private. Ferraresi et al. (2012) in their study reported that 'the implementation of effective Knowledge Management practices also depends on deliberate efforts to coordinate the diverse KM processes with the management of organizational resources as a whole, in alignment with a strategic vision' (p. 697).

Ding et al. (2013) in their research on knowledge transfer strategies reported that 'First, firms should choose their internal knowledge transfer strategies based on their knowledge storage forms, if firms mainly store knowledge using a systematization method, they should stress a codification strategy and encourage workers to access knowledge embodied in firms' documents, IT systems and so on. On the other hand, if firms' knowledge is mainly stored in employees' brains, firms should stress a rich-media strategy and encourage people to communicate and share knowledge. For example, firms should create culture promoting collaboration and develop places or facilities supporting workers contact' (p. 81). Lopez and Esteves (2013) conducted a research and reported that 'On one hand, companies apply the codification strategy, where knowledge is carefully codified and stored in an appropriate repository, enabling easy access and use by anyone in the Company. On the other hand, the personalization strategy should be applied when knowledge is closely tried to person who developed it and is shared mainly through direct person-to-person contacts' (p. 89). In view of the above, it is clear that strategy should be focusing around the people and consequently, it may be said that a suitable *strategy directs KM implementation favourably.*

2.5.3 Structure

While reviewing the literature, it was found that those organizations which are learning organizations are typically 'conceptualized as having a relatively flat structure, open communication systems, limited top-down control, and autonomous working conditions (Driver 2002). Handzic and Ozlen (2013) carried out a research on clinical service environments from Knowledge Management success and reported that the structures of 'both private and public healthcare organizations are increasingly implementing Knowledge Management solutions (KMS) to acquire, convert and provide access to relevant information and knowledge' (p. 1350011-1). In an exhaustive study on Knowledge Management, Hislop (2013) discussed the 'hypertext organization' and argued that this form of organization facilitates knowledge creation by synthesizing the efficiency achievable in hierarchical organizations with the adaptability of more flexible organizational structures (p. 106). In a case study, Kase et al. (2009) found that the design of work processes which encourage and facilitate collaboration (such as team working, job rotation), and team-/group-based pay which encourages/rewards people on the basis of group performance will provide a suitable structure for a Company. Husted et al. (2012) had conducted an empirical test of hostility and governance mechanisms and concluded that the governance mechanisms range from organizational forms such as partnership (Felin et al. 2009), project and organizational structure in hierarchies, communities and networks (Foss and Michailova 2009), work designs, training and development programmes, compensation systems, socialization techniques (Cabrera et al. 2006; Husted and Michailova 2009; Minbaeva and Pedersen 2010; Lopez-Cebrates et al. 2009; Simonin and Ozsomer 2009), and identity and identification (Argote and Kane 2009). The mechanisms are deployed with the expectations that influencing the conditions of individual actions in a certain manner will lead those individuals to make decisions that, when aggregated, lead to favourable organizational outcomes (Foss 2007). Mueller (2012) conducted a research on knowledge sharing between project teams and their cultural antecedents and described project teams, their departments and networks to show that they enable individual knowledge sharing. Therefore, it is concluded that *team-based matrix structure affects Knowledge Management Implementation.*

2.5.4 *Systems*

Organizational capital comprises of systems, processes and systemic routines. While reviewing the literature, it came to the fore that taking into account the effect of the organizational variables (i.e. delegation and extrinsic/intrinsic rewards) on knowledge transfer, organizational design has a significant impact in the achievement of organizational mission and, thus, on the efficiency of the organization (Matrin-Péres et al. 2012). In effect, the level of efficiency determines the survival of all organizations. The survival of the organization is largely dependent on the robust systems. Andries and Wastyn (2012) undertook a research in value-enhancing and cost-increasing effects of Knowledge Management and found out that the systems should include the tools and drivers of a written Policy regarding KM: incentives for employees to share information within the Company; specific resources to detect and acquire knowledge outside the Company; a Policy to involve external experts from universities, research institutes or other companies in projects if necessary; and regular updating of internal databases or manuals regarding common practices, lessons learned or expert advice are required. To test the hypothesis, the authors used structural equation modelling with manifest variables comprising of research and development, procurement of relevant software, obtaining licences from external sources, investing in the training of personnel to operationalise the licences, developing activities honed for introducing new products in the market and performing other activities for the implementation of new services and products to the market. The authors concluded that 'firms increasingly adopt KM techniques to spur the generation and acquisition of new knowledge by adopting the systems which can ensure a sustainable competitive advantage by putting tenable systems in place' (pp. 387–397). According to the outcome of review of the literature, it may be highlighted that *well-laid systems enable KM implementation suitably.*

2.5.5 *Processes*

The review of the literature has resulted in enumeration of a number of processes related with Knowledge Management. Gupta and Govindrajan (2000) while studying a case of Nucor Steel, talked about *creating* as KM process in order to hold the new knowledge inside the firm for exploration of new knowledge and its various recombinations. The process of creating knowledge was also brought out by Miller et al. (2007) while studying the patents in the realm of intellectual capital. Lenox and King (2004) propounded *integrating* as a KM process while studying the information and communication industries with a purpose of recognizing sources of knowledge and absorbing and integrating the knowledge within the firm. The authors specified the utilization of existing resources to generate economic rents' indefensible configurations and found that it is very important. Knowledge process comes into existence because of integration of groups and lateral and vertical socialization of groups. In addition to this, *reconfiguring* was reported as a result of study of Canon's AE-1 camera and the study of examination of Yahoo, which established itself as a Knowledge Management Process for use of familiar components and decrease of knowledge due to uncertainty in innovation. It was having the purpose to combine and deploy existing resources within the firm to reduce a competitive advantage (Denford 2013; Rindova and Kotha 2001). *Replicating* was established as another Knowledge Management Process to recognize, assimilate and apply existing resources elsewhere within the organization to ensure organization growth based upon knowledge redeployment. The process of replicating was reported by Winter and Szulanski (2001) as a result of the study of Banc One.

Dyer and Nobeoka (2000) studied Toyota and survey of strategic alliances conducted by Heimeriks and Duyster (2007) who propounded *developing* as a Knowledge Management Process because this could provide the benefit of mutual firm learning, resulting in new knowledge besides generating new knowledge outside the firm through recombination of firm and partner knowledge. *Assimilating* was discovered as a Knowledge Management Process which could serve for information outside the firm to absorb the knowledge into the firm and apply it to commercial means. It was having the benefit of growth of the firm through acquisition or inclusion of knowledge into industry networks (Lane and Lubatkin 1998). Macpherson et al. (2004) studied the supply network of British firms and found that the leveraging partners' resources to better exploit own resources in new configurations with the purpose of combining and redeploying existing firms and partner knowledge to create competitive advantage could be possible by *synthesizing*. The 'people' angle was brought into the Knowledge Management Processes by Denford (2013) while studying the Hollywood motion picture industry, when he used the word *imitating*. This was earlier reinforced by Cockburn et al. (2000), when they studied pharmaceutical industry. Thus, it may be inferred that *predefined processes help Knowledge Management Implementation.*

2.5.6 Management Support

Management support can be studied on two levels: (i) as a catalyst to Knowledge Management as part of change management and (ii) by providing the conducive leadership for implementation of Knowledge Management. The first role was studied by Kamhawy (2012), and he concluded his research by pointing out that without the help of the management the chance to 'work on only few ones and postpone some of the efforts and use of resources on other areas that KM needs, to some future stages may not work. Second, it also gives the management the opportunity to divide their whole KM-based change program into more consistent and smaller programs or projects. This may make the KM-based change programs more manageable and effective' (p. 820). The members of the management team are key informants for Knowledge Management purposes, and managers can be those managers in the organization that have access to, and use of, the organization's knowledge. This can be virtually possible to have KM by the help of managers in the organization (Gold et al. 2001). The role of middle managers is to take the high-level vision of organizational leaders and translate it into concepts and frameworks that are relevant to the workers they are responsible for. Thus, they are not passive conduits for passing on the ideas of leaders, but actively translate them into different terms. 'Middle managers have responsibility for communicating this translated vision to workers and motivating them to create knowledge in pursuit of it' (Hislop 2013, p. 114).

The second level of the management support is that of the leadership because the senior management team holds a position of respect among members of the organization, which provides objectives to the pursuant and also provides a sound judgmental trait without showing any favouritism to any special group (Fullwood et al. 2013). It has also been reported that the top management can use HRM practices to facilitate Knowledge Management initiatives by dealing with the problems and challenges that can often make workers unwilling to participate in Knowledge Management activities. In the study of leadership undertaken by Lee (2011), one of the ways team leaders developed intra-team trust and knowledge sharing was via a process of mentoring that involves linking experienced team members with less experienced ones. Thus, setting up of facilitating the establishment of both coaching and mentoring activities represents another way for organizational management to facilitate interpersonal knowledge sharing. Consequently, it proves that *management support is a prerequisite for Knowledge Management.*

2.5.7 Training

Training and development are the two areas which have got bearing on Knowledge Management, especially its implementation aspect. Jones and Mehon (2012) had studied the knowledge transfer in high velocity and turbulent environment and found that training is the recourse that can be taken to sell the Knowledge Management Implementation in an organization. They finally reported that in high-velocity environment, good training helps people work with ambiguity in rapidly changing environments as well as how to access the information and knowledge needed in real time with different communication methods. Training also helps people adapt in many different situations, with contingency planning, shared processes and shared understanding of situation and outcomes. Continual training and simulation also help people to anticipate and share knowledge. Other dimensions include the need to create social networks, develop a learning organization, develop knowledge maps to find the expertise needed plus mentoring programmes and understand social networks where the culture supports and facilitates multiple learning in collaborative social networks.

The training should be taken on organizational level so that the institutionalization at organization level of changes in behaviour could be ensured. Bui and

Baruch (2011) identified particular antecedents linked to each of these disciplines, with, for example, the antecedents of team learning being suggested as being team commitment, leadership, goal setting, development and training, organizational culture and individual learning. The training can be facilitated via the creation of 'learning cultures', where learning, reflection, debate and discussion are encouraged (Lopez et al. 2004; Raz and Fadlon 2006); the embedding of learning opportunities in organizational decision-making process could be possible (Carroll et al. 2006); project-based work is common, via processes such as post-project reviews (Ron et al. 2006; von Zedtwitz 2002). As a part of this discussion, self-development opportunity for all also included because they provide self-development opportunity to all staff to develop themselves as they see appropriate. During the review of literature, we found that the enormous literature has been produced on 'learning since mid-1990s and they are of great relevance to understand the dynamics of organizational knowledge processes' (Hislop 2013, p. 98). Consequently, it appears that *relevant training to personnel is a must for Knowledge Management Implementation*.

2.5.8 Reward and Recognition

The review of the literature regarding reward and recognition has brought to light the extrinsic and intrinsic motivation also. Kamhawy (2012) has concluded that there is a positive relation between Knowledge Management Implementation, and reward and recognition. Availability of proper reward systems is one of these non-IT-related factors (Kulkarni et al. 2007). It is claimed that such systems can be used to promote, at least in the beginning; applying KM initiatives, employees' levels of cooperation and collaboration, which are crucial aspects for Knowledge Management Awards in organizations (Bennett and Gabriel 1999; Goh 2002; Lee et al. 2005). It requires that the rewards and recognition should be structured in such a way that employees feel motivated towards Knowledge Management efforts of an organization. In talking about motivation, it is necessary to distinguish between intrinsic and extrinsic motivation. Intrinsic motivation refers to the pleasures and positive feelings people can derive from simply carrying out a task or activity, rather than for any reward derived from doing so (Hislop 2013, p. 222).

Quigley et al. (2007) stress the necessity of consideration of both intrinsic and extrinsic rewards when addressing the issue of knowledge transfer. Though extrinsic rewards help stability and continuity of the work force, it also has been found that monetary remuneration and perquisites given to the personal may faciliate motivation besides salary. The research has reported that recognition provides intrinsic reward and motivation. Intrinsic rewards are, therefore, powerful tools to overcome knowledge transfer barriers. Specifically, intrinsic rewards enable the development of informal group outside formal organizational structures, which allows rapid problem-solving, the transfer of improved practices and the creation of professional abilities. Furthermore, intrinsic rewards promote a working environment that expedites both formal and informal communication, which entails stronger organizational learning behaviours. Intrinsic rewards also may increase employees' commitment to the organization by creating self-improvement desires as a means to support the organization, bringing about the development of 'learn-to-learn' capabilities (Matrin-Péres 2012). Thus, intrinsic rewards perform two significant roles in the knowledge transfer process. First, they contribute to and promote employee participation in the knowledge transfer process. Second, they are a natural by-product generated by the process itself (Locus and Ogalivie 2006). From the above, it may be deduced that *rewards and recognition are motivators for effective Knowledge Management.*

2.6 Information and Communication Technology and Tools

The tools enabled by the information and communication technology and the information technology itself have largely been researched by scholars. However initially, it was reported that the management of knowledge tends to happen in an informal way 'rarely supported by purposely designed ICT systems' (Nunes et al. 2006). However, the researchers had investigated this issue further. IT strategies can be classified into two general categories: IT environment scanning and strategic use of IT (Bergeron et al. 2004). System KM strategy requires IT tools that allow for explicit knowledge to be formalized and articulated in documents, and shared electronically through IT infrastructures such as intranet. It has also been concluded by researchers that organizational culture influences not only the types of knowledge that should be transferred, e.g. norms of communication and knowledge sharing, (Zaidman and Brock 2009) but also attitudes towards using IT to share knowledge and advocate the creation of organizational cultures that encourage employees to use IT for knowledge sharing.

Knowledge can be shared via several means such as meetings, manuals, telephone conversations, seminars, conferences and on-the-job training. The intention to share knowledge should thus drive not only the use of IT to share knowledge but also the use of other means they may, in some instances, be more efficient than IT. It is a major enabler of Knowledge Management and a powerful means for sharing knowledge (Mitchell 2003). IT such as intranets, databases, e-mail, Web pages, bulletin boards and electronic forums provides effective knowledge-sharing mediums (Song 2001). Some of the scholars had focused on KM mainly as a technology issue (Davenport et al. 2008; Delmonte and Aronson 2004; Malhotra 2005), and the heavy emphasis on knowledge as a thing and/or process. By 2013, the definition of information and communication technology was codified by Hislop (2013) as ICTs are 'technologies which allow/facilitate the management and/or sharing of knowledge and information, thus, the term covers an enormous diversity of heterogeneous technologies including computers, telephones, e-mail, databases, data-mining systems, search engines, the internet, and video-conferencing equipment' (p. 203).

Therefore, it can be concluded that ICT is an important factor and it contributes to the Knowledge Management as an enabler. It is expected that *information and communication technology is an enabler for KM implementation.*

In addition to information and communication technology, various tools enabled by ICT also help implementation of Knowledge Management and they are intranets, data warehouses, decision support tools and groupware (i.e. technologies that support collaboration and communications). Scarbrough and Swan (2001) found that this emphasis was also reflected in academic research on Knowledge Management, with the vast majority of published search focusing on information technology-related issues. Since the mid-2000s, there has been a significant amount of research on this broad topic (Chiravuri et al. 2011; Ho et al. 2011; Faraj et al. 2011). In addition to these 'instant messenger' or 'chat' and expert database, corporate yellow pages (Directory of Subject Matter Experts), 'Blogging' has also been reported as KM tools for implementation of Knowledge Management (Matthews 2000). The review of the literature has brought in focus the Knowledge Map as an important tool which may be enabled by ICT or may not be enabled by it, but which helps Knowledge Management Implementation in an organization (Tiwana 2000). Since KM tools enabled or not enabled by ICT have also been reported as important drivers for Knowledge Management Implementation, it may be summarized that *tools related with communication are helpful for effective KM.*

2.7 Research Model Development

Having completed the review of the literature from the point of view of Knowledge Management, especially the implementation, it can be seen that culture is all pervasive and it is the most important factor in Knowledge Management and rest of the factors are having interface with culture. Not only this may 'fit' with the culture, it will be certainly successful. Therefore, the research model depicts culture as an influencing factor of Knowledge Management, overpowering every other factor. Thus, culture has been given the superimposition on every other factor. It has also been found that human resource management tools are mostly required rather than the tools provided by information and communication technology. This model has been further elaborated subsequently in research methodology chapter.

References

Afouni, F.: Human resource management and knowledge management: a road map toward improving organizational performance. J. Am. Acad. Bus. **11**, 124–131 (2007)

Akhavan, P., Jafari, M.: Critical issues for knowledge management implementation at a national level. J. Inf. Knowl. Manag. Syst. **36**(1), 52–66 (2006)

Al-Busaidi, K., Olfman, L.: An investigation of the determinants of knowledge management systems success in Omani organizations. J. Global Inf. Technol. Manag. **8**(3), 6–27 (2005)

Andrews, K.M., Delahaye, B.L.: Influences on knowledge processes on organizational learning: the psychosocial filter. J. Manag. Stud. **37**(6), 797–809 (2000)
Andries, P., Wastyn, A.: Disentangling value-enhancing and cost-increasing effects fof knowledge management. J. Knowl. Manag. **16**(3), 387–399 (2012)
Argote, L., Kane, A.A.: Superordinate identity and knowledge creation and transfer in organizations. In: Foss, N.J., Michailova, S. (eds.) Knowledge Governance: Processes and Perspectives, pp. 166–190. Oxford University Press, New York, NY (2009)
Asoh, D.A.: Business and knowledge strategies: alignment and performance impact analysis, Ph.D. dissertation. State University of New York (2004)
Audi, R.: Epistemology, a Contemporary Introduction to the Theory of Knowledge. Routledge, London (2003)
Bennett, R., Gabriel, H.: Organizational factors and knowledge management within large marketing departments: an empirical study. J. Knowl. Manag. **3**(3), 212–225 (1999)
Bergeron, F., Raymond, L., Rivard, S.: Ideal patterns of strategic alignment and business performance. Inf. Manag. **41**(1), 51–60 (2004)
Bergmann, J.R.: Ethonomethodology. In: Flick, et. al. (eds.) A companion to Quantitative Research, pp. 72–80. Sage Publications, London, (2004)
Berman, S.L., Down, J., Hill, C.W.L.: Tacit knowledge as a source of competitive advantage in the national basketball association. Acad. Manag. J. **45**(1), 3–31 (2002)
Bock, G.-W., Zmud, R.W., Kin, Y.-G.: Behavioral intention formation in knowledge sharing: examining the roles of extrinsic motivators, social-psychological forces and organizational climate. MIS Q. **29**(1), 11–87 (2005)
Bogner, W., Bansal, P.: Knowledge management as the basis of sustained high performance. J. Manag. Stud. **44**(1), 165–188 (2007)
Bollinger, A.S., Smith, R.D.: Managing organizational knowledge as a strategic asset. J. Knowl. Manag. **5**(1), 8–18 (2001)
Bowman, B.J.: Building knowledge management system. Inf. Syst. Manag. **19**(3), 32 (2002)
Browaeys, M.J.: Complexity of Epistemology: Theory of Knowledge or Philosophy of Science, In: 4th Annual Meeting of the European Chaos and Complexity in Organizations' Network; 22–23 October, 2004, Driebergan, nal (2004)
Brown, J.S., Duguid, P.: Knowledge and organisation: a social-practice perspective. Organ. Sci. **12** (2), 198–213 (2001)
Buehl, M.M., Alexander, P.A.: Motivation and performance differences in students, domain-specific epistemological beliefs profiles. Am. Educ. Res. J. Winter 2005, **42**(4), 697–726 (2005)
Bui, H., Baruch, Y.: Learning organization in higher education: an empirical evaluation within an international context, management learning (Publ. via Onlinefirst) (2011). doi:10.1177/1350507611431212
Cabrera, A., Colins, W.C., Salgado, J.F.: Determinants of individual engagement in knowledge sharing. Int. J. Hum. Resour. Manag. **16**(5), 720–735 (2006)
Carroll, J., Hatakenaka, S., Rudolph, J.: Naturalisitc decision making and organizational learning in nuclear power plants: negotiating meaning between managers and problem investigation teams. Organ. Stud. **27**(7), 1037–1057 (2006)
Casimir, G., Lee, K., Loon, M.: Knowledge sharing; influences of trust, commitment and cost. J. Knowl. Manag. **16**(5), 741–742 (2012)
Castro, G.M., Lopez-Sapez, P., Delgado-Verde, M.: Towards a knowledge-based view of firm innovation: theory and empirical approach. J. Knowl. Manag. **15**(6), 871–874 (2011)
Chatti, M.A.: Knowledge management: a personal knowledge network perspective. J. Knowl. Manag. **16**(5), 834 (2012)
Chawla, D., Joshi, H.: Knowledge management initiatives in Indian and public sector organisations. J. Knowl. Manag. **14**(6), 811–824 (2010)
Chen, Y.Y., Yeh, S.P., Huang, H.L.: Does knowledge management "fit" matter to business performance? J. Knowl. Manag. **16**(5), 671–687 (2012)

Chiravuri, A., Nazareth, D., Ramamurthy, K.: Cognitive conflict and consensus generation in virtual team during knowledge capture: comparative effectiveness of techniques. J. Manag. Inf. Syst. **28**(1), 311–350 (2011)
Cockburn, I.M., Henderson, R.M., Stern, S.: Untangling the origins of competitive advantage. Strateg. Manag. J. **21**(10/11), 1123–1145 (2000)
Cong, X., Pandya, K.V.: Issues of knowledge management in the public sector. Electron. J. Knowl. Manag. **1**(2), 25–33 (2003)
Coyte, R., Ricceri, F., Guthrie, J.: The management of knowledge resources in SMEs: an Australian case study. J. Knowl. Manag. **16**(5), 789–807 (2012)
Crane, L.: Trust me, I'm an expert: identity construction and knowledge sharing. J. Knowl. Manag. **16**(3), 457 (2012)
Cuervo-Cazurra, A., Un, A.: Why some firms never invest in formal R&D. Strateg. Manag. J. **31**, 759–779 (2010)
Dalkir, K.: Knowledge management in theory and practice, 2nd edn. The MIT Press, Cambridge, M.A. and London (2011)
Damodaran, L., Olphert, W.: Barrriers and facilitators to be use of knowledge management systems. Behav. Inf. Technol. **19**(6), 405–413 (2000)
Davenport, T., Klahr, P.: Managing customer support knowledge. Calif. Manag. Rev. **40**(3), 195–208 (1998)
Davenport, T.H., Prusak, L.: Working Knowledge, How Organizations, Manage What They Know. Harvard Business School Press, Boston, Massachusetts (1998)
Davenport, T., Prusak, L., Strong, B.: Putting ideas to work, MIT sloan management review. http://sloanreview.mir.edu/businessinsight/articles/2008/1/5011/ (2008)
Delmonte, A., Aronson, J.: The relationship between social interaction and knowledge management system success. J. Knowl. Manag. Pract. **5**, www.tlainc.com/articl71.htm (2004)
Denford, J.S.: Building knowledge:developing a knowledge-based dynamic capabilities typology. J. Knowl. Manag. **14**(2), 175–195 (2013)
Ding, X.H., Liu, H., Song, Y.: Are internal knowledge transfer strategies double-edged swords? J. Knowl. Manag. **17**(1), 81 (2013)
Doherty, N.F., Doig, G.: An analysis of the anticipated cultural impacts of the implementation of data warehouses. IEEE Trans. Eng. Manag. **50**(1), 77–88 (2003)
Donate, M., Guadamilas, F.: The effect of organizational culture of knowledge management practices and innovation. Knowl. Process Manag. **17**(2), 82–94 (2010)
Dredze, M., et. al.: Entity disambiguation for knowledge based population. Wikipedia accessed by using keyword 'Knowledge as an Entity' on 04.01.2013 (2013)
Driver, M.: The learning organization: Foucauldian gloom or Utopian sunshine? Hum. Relat. **55**(1), 33–53 (2002)
Dufour, Y., Steane, P.: Implementing knowledge management: a more robust model. J. Knowl. Manag. **11**(6), 68–78 (2007)
Dyer, J.H., Nobeoka, K.: Creating and managing a high performance knowledge-sharing network: the Toyota case. Strateg. Manag. J. **21**(3), 345–367 (2000)
Earl, M.: Knowledge management strategies: towards a taxonomy. J. Manag. Inf. Syst. **18**(1), 215–233 (2001)
Edgar, A., Sedgwick, P.: Cultural Theory, The Key Concepts. Routledge, Landon (2003)
Elliott, R.K., Jacobson, P.D.: The evolution of the knowledge professional. Acc. Horiz. **16**(1), 69 (2002)
Empson, L.: Introduction: knowledge management in professional service firms. Hum. Relat. **54**(7), 811–817 (2001)
Fairclough, N.: The discourse of new labour: critical discourse analysis. In: Yates, S., Taylor, S., Wetherell, M. (eds.) Discourse as Data: A Guide for analysis. Sage, London (2001)
Faraj, S., Javenpaa, S., Mazchrzak, A.: Knowledge collaboration in online communities. Organ. Sci. **22**(5), 1124–1139 (2011)

Felin, T., Zenger, T., Tomsik, J.: the knowledge economy: emerging organizational forms, missing microfoundations, and key considerations for managing human capital. Hum. Resour. Manag. **48**(4), 555–570 (2009)

Ferraresi, A.A., Quandt, C.O., dos Santos, S.A., Frega, R.J.: Knowledge management and strategic orientation: leveraging innovativeness and performance. J. Knowl. Manag. **16**(5), 688–701 (2012)

Flick, U.: Constructivism. In: Flick, et.al. (eds.) A companion to Quantitative Research, pp. 72–80. Sage Publications, London (2004)

Foss, N.J.: The emerging knowledge governance approach: challenges and characteristics. Organization **14**(1), 29–52 (2007)

Foss, N.J., Michailova, S.: Knowledge Governance Approach. Oxford University Press, New York, NY (2009)

Frame, D.: What is epistemology?: the theory of knowledge-is knowledge possible? http://doug-frame-suite101.com/whatisepistemology-a18352 (2010)

Fullwood, R., Rowley, J., Delbridge, R.: Knowledge sharing amongt academics in UK universities. J. Knowl. Manag. **17**(1), 123–136 (2013)

Goel, A.K., Sharma, G.R., Rastogi, R.: Knowledge management implementation in NTPC: an Indian PSU. Manag. Decis. **48**(3), 383–395 (2010)

Goh, S.C.: Managing effective knowledge transfer: anintegrative framework and some practice implications. J. Knowl. Manag. **6**(1), 23–30 (2002)

Gold, A.H., Malhotra, A., Segars, A.H.: Knowledge management: an organisational capabilities perspective. J. Manag. Inf. Syst. **18**(1), 185–214 (2001)

Gupta, A.K., Govindrajan, V.: Knowledge management's social dimension: lessons from Nucor steel. Sloan Manag. Rev. **42**(1), 71–80 (2000)

Haas, M., Hansen, M.: Different knowledge, different benefits: towards a productivity perspective on knowledge sharing in organizations. Strateg. Manag. J. **28**, 1133–1153 (2007)

Han, X., Sun, L.: A generative entity model for linking entities with knowledge base. Hum. Lang. Technol. **I**, 945–954 (2011)

Handzic, M., Ozlen, K.: Knowledge management success in clinical service environments. J. Inf. Knowl. Manag. **12**(2), 1350011-1 (2013)

Harvey, J.F.: Managing organizational memory with intergenerational knowledge transfer. J. Knowl. Manag. **16**(3), 400–417 (2012)

Heimeriks, K.H., Duysters, G.: Alliance capability as a mediator between experience and alliance performance: an empirical investigation into the alliance capability development process. J. Manag. Stud. **44**(1), 24–29 (2007)

Hislop, D.: Knowledge Management in Organizations: a Critical Introduction. Oxford University Press, Oxford (2013)

Ho, S.-C., Ting, P.-H., Bau, D.-Y., Wei, C.-C.: Knowledge sharing intention in a virtual community: a study of participants in the Chinese wikipedia. Cyber Psychol. Behav. Soc. Netw. **14**(9), 541–545 (2011)

Hofer, B.K., Pintrich, P.R.: Personnel Epistemology: The Psychology of Beliefs About Knowledge and Knowing. Erlbaum, Mahwah, NJ (2002)

Holsapple, C.W., Joshi, K.D.: The inseparability of modern knowledge management and computer-based technology. J. Knowl. Manag. **9**(1), 42–52 (2001)

Huang, J.C., Newell, S., Galliers, R.D., Pan, S.: Dangerous liaisons? Component-based development and organizational subcultures. IEEE Trans. Eng. Manag. **50**(1), 89–99 (2003)

Husted, K., Michailova, S.: Socialization tactics as a governance mechanism in R&D collaborations. In: Foss, N.J., Michailova, S. (eds.) Knowledge Governance: processes and Perspectives, pp. 191–219. Oxford University Press, New York, NY (2009)

Husted, K., Michailova, S., Minbeava, D.B., Pedersen, T.: Knowledge-sharing and governance mechanisms an empirical test. J. Knowl. Manag. **16**(5), 754–773 (2012)

Ipe, M.: Knowledge sharing in organizations: a conceptual framework. Human Res. Dev. Rev. **2**(4), 337–359 (2003)

Jones, N.B., Mahon, J.F.: Nimble knowledge transfer in high velocity/turbulent environments. J. Knowl. Manag. **16**(5), 724–788 (2012)
Kamhawy, E.M.: Knowledge management fishbone: a standard framework of organizational enablers. J. Knowl. Manag. **16**(5), 808–828 (2012)
Kase, R., Paauwe, J., Zupan, N.: HR practices interpersonal relations and intra-firm knowledge transfer in knowledge intensive firms: a social network perspective. Hum. Resour. Manag. **48**(4), 615–639 (2009)
King, W., Marks, P.: Motivating knowledge sharing through a knowledge management system. Omega **36**, 131–146 (2008)
Koenig, M.E.D., Jank, D.A.: The (common) sense of KM. J. Inf. Knowl. Manag. **11**(2), 1250009-1–1250009-5 (2012)
Kulkarni, U.R., Ravindran, S., Freeze, R.: A Knowledge management success model: theoretical development and empirical validation. J. Manag. Inf. Syst. **23**(3), 309–347 (2007)
Lane, P.J., Lubatkin, M.: Relative absorptive capacity and interoganizational learning. Strateg. Manag. J. **19**(5), 461–477 (1998)
Lee, K.C., Lee, S., Kang, I.W.: KMP: measuring knowledge management performance. Inf. Manag. **42**, 469–482 (2005)
Lee, L.: The effects of challenge and hinderance on unlearning and NDP success: the moderating role of team conflict. Afr. J. Bus. Manag. **5**(5), 1843–1856 (2011)
Lenox, N., King, A.: Prospects for developing absorptive capacity through internal information provision. Strateg. Manag. J. **25**(4), 331–345 (2004)
Levin, D.Z., Cross, R.: The strength of weak ties you can trust: the mediating role of trust in effective knowledge transfer. Manag. Sci. **55**(11), 1477–1490 (2004)
Locus, L.M., Ogalivie, D.: Things are not always what they seem: how reputations, culture and incentives influence knowledge transfer. Learn. Organ. **13**(1), 7–24 (2006)
Lopez, S.P., Peon, J.M.M., Ordas, C.J.V.: Managing knowledge: the link between culture and organizational learning. J. Knowl. Manag. **8**(6), 93–104 (2004)
Lopez, V.W.B., Esteves, J.: Acquiring external knowledge to avoid re-invention. J. Knowl. Manag. **17**(1), 89 (2013)
Lopez-Cebrates, A., Prez-Luo, A., Cabrera, R.V.: Knowledge as a mediator between HRM practices and innovative activity. Hum. Resour. Manag. **48**(4), 485–503 (2009)
Macpherson, A., Jones, O., Zhang, M.: Evolution or revolution? Dynamic capabilities in a knowledge dependent firm. R&D Manag. **34**(2), 161–177 (2004)
Malhotra, Y.: Integrating knowledge management technologies in organizational business processes: getting real time enterprises to deliver real business performance. J. Knowl. Manag. **9**(1), 7–28 (2005)
Mason, D., Pauleen, D.J.: Perceptions of knowledge management: a qualitative analysis. J. Knowl. Manag. **7**(4), 38–48 (2003)
Matrin-Péres, V., Martín-Cruz, N., Estrada-Vaquero, I.: The influence of organizational design on knowledge transfer. J. Knowl. Manag. **16**(3), 418–434 (2012)
Matthews, M.L.: Knowledge-Driven Profit Improvement: Implementing Assessment Feedback Using PDK Action Theory. Luice Press, London, New York, St (2000)
Meuller, J.: Knowledge sharing between project teams and its cultural antecedents. J. Knowl. Manag. **16**(3), 435–447 (2012)
Miller, D.J., Fern, M.J., Cardinal, L.B.: The use of knowledge for technological innovation withing diversified firms. Acad. Manag. J. **50**(2), 308–322 (2007)
Minbaeva, D., Pedersen, T.: What drives knowledge sharing behavior of individuals? Int. J. Strateg. Change Manag. **22**(3), 200–222 (2010)
Mitchell, H.: Technology and knowledge management: is technology just an enabler or does it also add value? In: Coakes, E. (ed.) Knowledge Management: Current Issues and Challenges, pp. 66–78. IRM Press, London (2003)
Mohamed, M.S., O'Sullivan, K.J., Ribiere, V.: A paradigm shift in the Arab region knowledge evolution. J. Knowl. Manag. **12**(5), 107–120 (2008)

Nag, R., Gioia, D.: From common to uncommon knowledge: foundations of firm-specific use of knowledge as a resource. Acad. Manag. J. **55**(2), 421–457 (2012)

Nonaka, I., von Krogh, G., Voelpel, S.: Organizational knowledge creation theory: evolutionary paths and future advances. Organ. Stud. **27**(8), 1179–1208 (2006)

Nonaka, I., Toyana, R., Nagata, A.: Knowledge entity. Ind. Corp. J. **9**, 1–20 (2012)

Nunes, M.B., Annansingh, F., Eaglestone, B., Wakefield, R.: Knowledge management issues in knowledge intensive SMEs. Knowl. Manag. Issues **62**(1), 101–119 (2006)

O'Dell, C., Elliott, S., Hubert, C.: Knowledge Management: A Guide for Your Journey to Knowledge Management Best-Practices. American Productivity & Quality Centre, Houston, TX (2000)

Pacharpha, T., Ractham, V.V.: Knowledge acquisition: the roles of perceived value of knowledge content and source. J. Knowl. Manag. **16**(5), 724–739 (2012)

Paramsothy, V., Woods, P., Raman, M.: Success factors for implementation of entrepreneurial knowledge management in Malaysian banks. J. Inf. Knowl. Manag. **12**(2), 1350015-1–1350015-19) (2013)

Quigley, N.R., Tesluk, P.E., Locke, E.A., Bartol, K.M.: Multilevel investigation of the motivational mechanisms underlying knowledge sharing and performance. Organ. Sci. **18**(1), 71–92 (2007)

Raz, A., Fadlon, J.: Managerial Culture, Workplace Culture and Situated Curricula in Organizational Learning. Organization Studies **27**(2), 165–182 (2006)

Renzi, B.: Trust in management and knowledge sharing–the mediating effects of fear and knowledge documentation. Omega-Int. J. Manag. Sci. **36**(2), 202–220 (2008)

Reus, T., Liu, Y.: Rhyme and reason: emotional capability and the performance of knowledge-intensive work group. Hum. Perform. **17**(2), 245–266 (2004)

Rindova, V.P., Kotha, S.: Continuous 'morphing' competing through dynamic capabilities form and function. Acad. Manag. J. **44**(6), 1263–1280 (2001)

Robey, D., Boudreau, M.C.: Accounting for the contradictory organizational consequences of information technology: theoretical directions and methodological implications. Inf. Syst. Res. **10**(2), 167–185 (1999)

Ron, N., Lipshitz, R., Popper, M.: How organizations learn: post-flight reviews in an F-16 fighter squadron. Organ. Stud. **27**(8), 1069–1089 (2006)

Ross, J., Weil, P.: Enterprise architecture as strategy: creating a foundation for business execution. Harvard Business School Press (2006)

Scarbrough, H., Swan, J.: Explaining the diffusion of knowledge management. Br. J. Manag. **12**(1), 3–12 (2001)

Shih, H.A., Chiang, Y.H.: Strategy alignment between HRM, KM and corporate development. Int. J. Manpower **26**(6), 582–606 (2005)

Siddique, C.M.: Knowledge management initiatives in the United Arab Emirates; a baseline study. J. Knowl. Manag. **16**(5), 702–703 (2012)

Simonin, B.L., Ozsomer, A.: Knowledge process and learning outcomes in MNCs: an empirical investigation of the role of HRM practices in foreign subsidiaries. Hum. Resour. Manag. **48**(4), 505–530 (2009)

Song, S.: An internet knowledge sharing system. J. Comput. Inform. Syst. **20**(2), 143–160 (2001)

Soo, C., Divenney, T., Midgley, D., Deering, A.: Knowledge management: philosophy, processes and pitfalls. Calif. Manag. Rev. **44**(4), 129 (2002)

Spencer, N.: http://ethicalpolitics.org/seminars/neville.htm (2012)

Stock, G., McFadden, K., Goven III, C.: Organizational culture, knowledge management, and patient safety in U.S. hospitals. Qual. Manag. J. **1**(2), 7–26 (2010)

Sullivan, D., Marvel, M.: Knowledge acquisition, network reliance, and early-stage technology venture outcomes. J. Manag. Stud. **48**(6), 1169–1193 (2011)

Sveiby, K.E.: A knowledge-based theory of the firm to guide in strategy formulation. J. Intellect. Capital **2**, 344–385 (2001)

Tamilin, A., Magnini, B., Serafini, L.: Leveraging entity linking by contextualized background knowledge. Wikipedia accessed by using keyword 'Knowledge as an Entity' on 30.12.2012 (2012)

Tan, B.C.Y., Smith, J., Keil, M., Montealegre, R.: Reporting bad news about software projects: impacts of organizational climate and information asymmetry in an individualistic and collectivistic culture. IEEE Trans. Eng. Manag. **50**(1), 64–77 (2003)

Tiwana, A.: The Knowledge Management Toolkit, Prentice-Hall International (U.K.) Limited, London (2000)

Tseng, C.-Y., Pai, D.C., Hung, C.-H.: Knowledge absorptive capacity and innovation performance in KIBS. J. Knowl. Manag. **15**(6), 971–983 (2011)

van Mujjen, J.J., Koopman, P., De Witte, K., De Cock, G., Susanj, Z., Lemoine, C.: Organizational culture: the focus questionnaire. Eur. J. Work Organ. Psychol. **8**(4), 551–568 (1999)

Veolpel, S., Dous, M., Davenport, T.: Five steps to creating a global knowledge-sharing systems: siemens' share net. Acad. Manag. Executive **19**(2), 9–23 (2005)

von Zedtwitz, M.: Organizational learning through post-project reviews in R&D. R&D Manag. **32**(3), 255–268 (2002)

Wang, D., Su, Z., Yang, D.: Organizational culture and knowledge creation capability. J. Knowl. Manag. **15**(3), 363–373 (2011)

Wang, H., He, J., Mahoney, J.: Firm-specific knowledge resources and competitive advantage: the role of economic-and relationship-based employee governance mechanism. Strateg. Manag. J. **30**, 1265–1285 (2009)

Webster, J., Brown, G., Zweig, D., Connelly, C.E., Brodt, S., Sitkin, S.: Beyond knowledge sharing: withholding knowledge at work. Res. Pers. Hum. Resour. Manag. **27**, 1–37 (2008)

Williams, C.: Client-vendor knowledge transfer in IS offshore outsourcing: insights from a survey of Indian software engineers. Inf. Syst. J. **21**, 335–356 (2011)

Winter, H.G., Szulanski, G.: Replication as strategy. Organ. Sci. **12**(6), 730–743 (2001)

Witherspoon, C.L., et al.: Antecedents of organizational knowledge sharing: meta-analysis and critique. J. Knowl. Manag. **17**(2), 250–277 (2013)

Xie, S., Markus H., Ostrowski L.: Information architectures for information sharing management —a literature review. J. Inf. Knowl. Manag. **11**(2), 125008-(1–12) (2012)

Zaidman, N., Brock, D.: Knowledge transfer within multinationals and their foreign subsidiaries: a culture-context approach. Group Organ. Manag. **34**(3), 297–329 (2009)

Chapter 3
The Description of the Case (POWERGRID)

3.1 The Description of the Case (POWERGRID)

Power Grid Corporation of India Limited (POWERGRID), during relatively short span of time, has emerged as a global leader in power transmission and earned admirable reputation of being one of the best managed utilities and technology leader in long-distance bulk power transmission. POWERGRID is a flag-bearer Company of the power sector discharging great responsibility of transmission of power from one part to any other part of the country with economy, reliability and security.

This Central Public Sector Enterprise of Ministry of Power, Government of India, is a designated 'Central Transmission Utility of the Country (CTU)'. Further, recognizing the contribution of POWERGRID in the overall development of power sector and based on its impeccable performance, its status has been upgraded to 'Navratna' in May 2008.

It has been listed in BSE and NSE after its initial public offer in 2007 and 'Follow on Offers' in 2010 and 2013. Presently, Govt. of India's holding is about 57.9 and 42.1 % is by Public.

Company owns and operates about 1,10,514 ckt kms of transmission lines at 800/765, 400, 220 and 132 kV EHVAC & +500 kV HVDC levels and 186 sub-stations. POWERGRID is undertaking grid management in the country through its fully owned subsidiary Company, namely Power System Operation Corporation Limited (POSOCO).

In order to create value to its stakeholders by utilizing its existing transmission infrastructure network and expertise, POWERGRID has successfully diversified in consultancy and telecom businesses. Its Site Offices are located internationally for carrying out the Company's overseas consultancy assignments.

POWERGRID with its brand name 'POWERTEL' is involved in telecom business and is having a telecom network of about 30,000 km and is one of the implementing agencies of the *National Knowledge Network project* (*NKN*) to

K.N. Pandey, *Paradigms of Knowledge Management*, Studies in Systems, Decision and Control 60, DOI 10.1007/978-81-322-2785-4_3

connect knowledge centres such as IITs, IIMs and NITs and the National Optical Fibre Network (NOFN) project of Government of India to connect 250,000 gram panchayats of the country.

Further, backed by its strong in-house expertise in various facets of transmission, sub-transmission, distribution and telecom sectors, the Company is executing various consultancy assignments domestically and internationally. In overseas operations, the Company is now working in 18 countries.

POWERGRID has also taken lead role in Govt. of India's national building schemes of Accelerated Power Development and Reforms Programme (APDRP) and Rajiv Gandhi GrameenVidyutikaranYojana (RGGVY). In addition, POWERGRID is playing a key role in the formation of SAARC grid, implementation of smart grid and prioritized 'Green Energy Corridors' in the country.

3.1.1 Formation of POWERGRID

Pursuant to the Government of India's Policy decision to form a national power grid to pave the way for the integrated operation of the central and regional transmission systems, POWERGRID was incorporated earlier as 'National Power Transmission Corporation Limited (NPTC)', a Government of India Enterprise on 23 October 1989 under the Companies Act 1956 with the responsibility of planning, executing, owning, operating and maintaining the high-voltage transmission systems. Subsequently, transmission assets from central generating companies and other Central/Joint Sector Organizations such as NTPC, NHPC, NEEPCO, NLC, NPC, THDC, SJVNL (formerly NJPC) were transferred to POWERGRID in a phased manner and it commenced commercial operation in 1992–93. In addition, Regional Load Despatch Centres (RLDCs) from Central Electricity Authority were transferred to POWERGRID in a phased manner from 1994 to 1996. These RLDCs were modernized progressively and National Load Despatch Centre (NLDC) was established in 2009 for overall coordination at National level.

3.1.2 National Grid

Recognizing the need for the development of National Grid, thrust was given to enhance the capacity of inter-regional links in a phased manner. In this direction, various inter-regional links were established/planned as a part of different generation projects/grid strengthening schemes. Towards this, inter-regional power transfer capacity of the National Grid has been enhanced to about 44,250 MW.

All India synchronous grids have become a reality on 31 December 2013 for optimal utilization of the resources in the country. With this, 'One Nation-One Grid-One Frequency' has become operational. Towards this presently, regional grids of Northern, Eastern, Western and North-eastern regions (NEW grid) are already connected through synchronous interconnections and Southern region (SR) is connected to this NEW grid through over 4000 MW capacity HVDC links. Synchronous interconnection of SR with NEW grid is through high capacity 765 kV Raichur–Sholapur lines (2 Nos.) for establishment of ‘All India Synchronous National Grid’ for facilitating bulk transfer of power across regional boundaries.

The Inter-regional power transfer capacity of the National Grid is planned to increase to about 66,000 MW at the end of XII Plan by adding about 38,400 MW during XII Plan matching with generation capacity addition programme and power transfer requirement across the regions.

National Grid has facilitated inter-regional power exchanges across the country. The increased inter-regional power exchanges have helped in meeting more demand in energy-deficit regions besides achieving overall economy.

Generation capacity addition of about 88,000 MW is envisaged during XII Plan, in which large numbers of generation projects are coming up in resource-rich states, i.e. Odisha, Jharkhand, Sikkim, Madhya Pradesh, Chhattisgarh, Tamil Nadu, and Andhra Pradesh under private sector. POWERGRID, the Central Transmission Utility and nodal agency for grant of long-term access (LTA) to private producers has undertaken development of High Capacity Transmission Corridors for the evacuation of large quantum of power from various independent power producers (IPP) projects. For integrated development of transmission system for dispersal of power from these generation projects, eleven (11) nos. of High Capacity Power Transmission Corridors (HCPTC) have been planned at an estimated cost of about 75,000 crore. Out of this, POWERGRID is to implement HCPTCs work of about 66,000 crore. Implementation of these corridors is taken up in phased manner matching with generation projects.

The major areas of concern in the development of transmission network in the country are conserving the right-of-way (RoW), minimizing impact on natural resources, coordinated development of cost-effective transmission corridor and flexibility in the upgradation of transfer capacity of lines.

To conserve RoW, POWERGRID is deploying technologies such as high-temperature low sag (HTLS) conductors, series compensation including thyristor control, multicircuits, compact and tall towers, high surge impedance loading lines. Further, in order to address these issues effectively and to ensure development of efficient, coordinated and economical inter-state transmission system, the Company has undertaken several state-of-the-art technological innovations. Company gives priority to research activities with potential for societal, environmental and national benefits by the application of advance technologies and finding solutions to gear up for future challenges.

Experienced with construction of 765 kV ultra-high-voltage AC (UHVAC) and ±500 kV HVDC transmission system, POWERGRID is now implementing on next higher transmission voltages of ±800 kV HVDC and developing 1200 kV UHVAC system to achieve efficient utilization of RoW and increased power transfer capability for transfer of bulk power over long distances.

3.1.2.1 ±800 kV Multiterminal HVDC System

POWERGRID is implementing ±800 kV, 6000 MW multiterminal HVDC system of around 2000 km from North-eastern region (Biswanath Chariali in Assam and Alipurduar of West Bengal) to Northern region (Agra in Uttar Pradesh) and upon completion, it shall be the one of the longest HVDC lines in the world.

3.1.2.2 1200 kV UHVAC Transmission System

To meet the long-term bulk power transfer requirement, an overlaying super grid with 1200 kV UHVAC system has been envisaged. The 1200 kV UHVAC technology, the highest voltage level in the world, is being developed by the Company in collaboration with 35 Indian manufacturers. This is one of the unique R&D projects in public–private partnership model. The pilot 1200 kV single circuit (S/c) and D/c lines were successfully test-charged at 1200 kV UHVAC National Test Station at Bina, Madhya Pradesh.

Further, technological initiatives such as pollution and lightning mapping of transmission system, process bus technology and superconducting (zero resistance) technology are being taken up/explored for their viability in Indian Power System to ensure more efficient, safe, secure and reliable operation of grid.

3.1.3 Operational Excellence

POWERGRID consistently maintains high availability of its transmission network by implementing technologically advanced operational techniques such as Hotline maintenance and hotline washing of insulators with the use of helicopters. Presently, 43 substations of POWERGRID are being operated remotely and many new substations are being designed for remote operation. For further improvement in operational efficiency, a ‘National Transmission Asset Management Centre (NTAMC)’ has being established to facilitate centralized operation, monitoring and control of substations remotely. ‘Maintenance service hub’ facilities are also created for optimal utilization of manpower and other resources of a group of substations and to reduce the response time in case of faults and breakdowns.

Average availability of transmission systems has been maintained consistently above 99 %, at par with international level. Number of trippings per line has been reduced to 0.58.

To meet eventualities caused by earthquake, fire, cyclone, landslides, sabotage, etc., the Company has kept state-of-the-art Emergency Restoration Systems (ERS) at strategic locations for restoration of collapsed transmission line towers in shortest possible time. Further, to mitigate the above eventualities in the case of substation, a mobile 400/220 kV ERS substation is being procured, which will facilitate quick deployment and restoration of substation in only 2–4 weeks, which otherwise may take several months.

3.1.4 Towards SAARC Grid

POWERGRID continues to play an active role in preparing a road map for developing South Asian Association for Regional Cooperation (SAARC) market for electricity to develop a cross-country power grid, harnessing each other's capacities and resources to address the growing energy need in the region. Presently, various interconnections exist between India and Nepal, and India and Bhutan, and these are being strengthened for mutual exchange of power. For the evacuation of power from various upcoming hydroelectric power (HEPs) in Bhutan, Punatsangchu-I HEP (in Bhutan)–Alipurduar (in India) 400 kV Double circuit (D/c) line between Bhutan and India has been completed by 2015. Further, an asynchronous interconnection between India and Bangladesh through 500 MW high-voltage direct current (HVDC) back-to-back terminal along with Bheramara (Bangladesh)–Baharampur (India) 400 kV D/c line is implemented. For transfer of bulk power, interconnection between India and Nepal through 400 kV Dhalkebar (in Nepal)–Muzaffarpur (in India) D/C transmission line is under implementation. For interconnection between India and Sri Lanka, feasibility study for a ±400 kV, 500/1000 MW under-sea HVDC bipole line is under finalization. Further, discussions at Government level are being held for interconnection between India and Pakistan through Amritsar (India)–Lahore (Pakistan) line.

3.1.5 Development of Electricity Market

POWERGRID, as Central Transmission Utility (CTU), is the nodal agency for processing and grant of connectivity, medium-term open access (MTOA) and long-term access (LTA) of various applicants. Till 31 March 2014, the Company granted LTA to 148 nos. of applications for 80,000 MW capacity, out of 223 applications received in all. Further, connectivity was granted for 79 nos. of eligible applications for 70,000 MW capacity out of 190 applications received for about 195,000 MW capacity in all. Till 31 March 2014, 127 applications were received for MTOA, and MTOA was granted to 48 nos. of applications in all for about 4,500 MW capacity.

Due to open access provided on inter-state transmission system, the electricity unit price has been coming down significantly.

3.1.6 Tariff-Based Competitive Bidding Regime

With effect from 6 January 2011 pursuant to the Tariff Policy and guidelines related thereto issued by the Government of India, all the Inter-State Transmission System (ISTS) Projects except for some specifically identified projects as may be determined by the Government of India are being executed through Tariff-based Competitive Bidding (TBCB). As per the Bidding guidelines, for each ISTS Project under TBCB, the Bidder who quotes the lowest levellized tariff has to acquire the Project SPV and establish the project on Build, Own, Operate and Maintain ('BOOM') basis. POWERGRID has since been participating in the Tariff-based Competitive Bidding.

Leveraging its techno-economical know-how, POWERGRID emerged as successful Bidder and won five transmission projects and acquired SPVs, viz. Vizag Transmission Limited, POWERGRID NM Transmission Limited, Unchahar Transmission Limited, NRSS XXXI (A) Transmission Limited and Vemagiri Transmission Limited. All projects are under various stages of implementation.

3.1.7 Contribution in Distribution Reforms

Govt. of India's 'Rajiv Gandhi GrameenVidyutikaranYojana (RGGVY)' scheme for Rural Electricity Infrastructure and Household Electrification was launched for the attainment of the National Common Minimum Programme of providing access to electricity to all rural households.

The Company has taken a lead role in the implementation of RGGVY works and has been assigned one-third of the total works for execution of rural electrification covered in 68 districts of 9 States in the country at a base cost of about Rs. 7,230 crore. Most of the schemes have already been completed. During FY 2013–14, infrastructure was created for electrification in 2388 villages out of which 125 were unelectrified villages. Service connections were provided to about 85,460 BPL households. Cumulatively, till March 2014, infrastructure has been created for electrification of 71,042 villages out of which 32,588 were unelectrified villages and service connection to about 36.22 lakh BPL households were provided.

3.1.8 Sustainable Development

Transmission projects are environmentally clean and involve neither any disposal of solid waste, effluents and hazardous substances in land, air and water nor large-scale excavation which may result in soil erosion. Transmission line projects rather help in reducing environmental impacts of power projects, especially thermal power projects. Generation Projects could be located faraway from inhabited areas to have minimum transportation of coal and for bulk power transfer, high-capacity transmission lines are set up to dispatch power to load centre located distance apart.

POWERGRID has proactively followed the Environmental and Social Policy and Procedures (ESPP) for the conservation of forests, flora and fauna, resettlement and rehabilitation. POWERGRID believes that its guiding principles of reliability, security and economy have to match up to the rising expectations of a cleaner, safer, healthier environment for people, both affected and benefited by its activities.

POWERGRID was the first Company in power sector to come out with 'Sustainability Report' and 2nd such report has been published in March 2013, providing all stakeholders a clear picture of contributions to sustainable development through its activities.

POWERGRID has taken various initiatives towards sustainable development and in the direction of reducing its carbon footprint. POWERGRID has adopted rain water harvesting facility in its establishments. Every substation being constructed now is being provided with rain water harvesting systems. Light-emitting diode (LED) bulbs and solar lights are being provided in new substations to reduce our energy requirement. Recently, a waste paper recycling plant has been commissioned at Gurgaon Substation. New buildings are being constructed as per Green Building norms.

Transfer of huge quantum of power across the regions poses massive challenges such as RoW and clearing of forest. Therefore, Company is focusing on more number of high capacity ±800 kV HVDC systems and 765 kV D/c lines to minimize RoW and environmental problems. In addition to Company is developing 1200 kV AC technology, highest AC voltage level in the world, as next higher AC voltage in India. The 1200 kV AC voltage line would enable transfer of 7000–8000 MW power over single corridor which will help in utilizing the right-of-way in an optimal manner and minimizing of loss of forest, flora and fauna, and scarce natural resource such as land.

3.1.9 Corporate Social Responsibility (CSR)

As a part of Corporate Social Responsibility, POWERGRID is making contribution to the society at large with emphasis on socio-economic and integral development of areas/communities primarily in and around its areas of operations and has carried out various community development activities such as skill development and capacity building, livelihood generation, health care, education, plantation, sanitation, drinking water, besides infrastructure developments such as classrooms, roads, community centres. During the FY 2013–14, POWERGRID made an expenditure of Rs. 21.66 crore on various CSR activities.

POWERGRID has also collaborated with number of specialized agencies to roll out its CSR initiatives. Some of the significant interventions were in the field of capacity building training to youths in 'Transmission Line Tower erection' which are being conducted in association with leading manufacturers. Approximately, 1200 persons have been trained through capacity building programme and gainfully employed in the transmission industry. Livelihood training programmes were also imparted to unemployed rural youth in tailoring, cutting/stitching, embroidery, automobile repairing, handloom weaving, food and fruit processing, repairing of household appliances, etc., to secure employment or be self-employed.

POWERGRID has completed Govt. of India's prestigious assignment Pul-e-Khumri–Kabul 220 kV Double circuit transmission line and 220/110/20 kV Chimtala Substation, in Kabul, Afghanistan in 2009 and power flow to Kabul has commenced from neighbouring country, Uzbekistan. The project was inaugurated by His Excellency Mr. Hamid Karzai, the President of Islamic Republic of Afghanistan in May 2009. It was one of the largest infrastructure projects ever undertaken in Afghanistan. Certificates of appreciation were given by Hon'ble Minister of Energy and Water, Afghanistan, in recognition of the difficulties and challenges involved in the project implementation and contributions made by POWERGRID.

3.1.10 POWERGRID in XII Plan

Considering the generation capacity addition of 88,000 MW envisaged during XII Plan, total fund requirement for transmission sector is estimated at about 180,000 crore. In this, an investment of about Rs. 100,000 crore is planned by POWERGRID during XII Plan for the further development of inter-state transmission systems which includes development of High Capacity Power Transmission Corridors (HCPTCs) apart from inter-regional links for enhancement of inter-regional power transfer capacity of National Grid and various system-strengthening schemes. During the XII plan, it has been envisaged to include about 40,000 ckm of transmission line and about 1,00,000 MVA of transformation capacity.

Keeping pace with changing business environment, POWERGRID had set its Vision and aligned its Mission and Objectives as a vanguard of foresight which are as follows:

3.1.10.1 Vision

World Class, Integrated, Global Transmission Company With Dominant Leadership in Emerging Power Markets Ensuring Reliability, Safety and Economy.

3.1.10.2 Mission

We will become a Global Transmission Company with Dominant Leadership in Emerging Power Markets with World Class Capabilities by:

- World Class: Setting superior standards in capital project management and operations for the industry and ourselves.
- Global: Leveraging capabilities to consistently generate maximum value for all stakeholders in India and in emerging and growing economies.
- Inspiring, nurturing and empowering the next generation of professionals.
- Achieving continuous improvements through innovation and state of the art technology.
- Committing to highest standards in health, safety, security and environment.

3.1.11 Technological Initiatives

Continuous upgradation in technology adapted for conservation of right-of-way (RoW), optimization of land for substations, improvement in the performance of transmission system, optimize costs leading to development of cost-effective transmission systems.

- POWERGRID has been continuously making R&D efforts such as design of tall, multicircuit and compact transmission line towers, use of high-temperature endurance conductors, remote-operated substations, compact substations layouts, which contribute towards reduction in transmission loss as a whole besides reduction in RoW for bulk power transmission.
- Power system availability of CTU's transmission network has been consistently over 99 %, comparable with the best international standards.
- Operating voltage level of the Indian Grid enhanced from 400 to 765 kV AC in October 2007.
- Work on next higher transmission voltages taken up to meet the long-term power transfer requirement in most effective manner.
- Implementation of ±800 kV, 6000 MW HVDC Bi-pole line from North-eastern region to Northern region has commenced in the country, which shall be the first of its kind (±800 kV HVDC line) having the largest power carrying capacity and transmitting power over distance of more than 2000 kms.
- A 1200 kV UHVAC Test Station along with a 1200 kV test line is being established at Bina in Madhya Pradesh by POWERGRID, as a collaborative effort with equipment manufacturers, for indigenous development of 1200 kV equipment. These technological initiatives are pioneering efforts even globally.
- Leadership role taken for implementation of smart grid in the country for most efficient grid operation.

All these measures facilitated development of efficient, coordinated and economical inter-state transmission system leading to cost reduction in transmission of power for valued customers.

In 2009–10, POWERGRID improved its ranking PLATTS fastest growing Asian Companies at 8th position and also among the fastest growing global energy companies at 18th position by PLATTS in their ranking of energy companies for 2010. First power sector enterprise which came up with Environment and Social Policy and Procedures Document in 1998 which was updated from time to time and the first CPSE in power sector to 'Sustainability Development'.

- Availability maintained consistently at >99 % without any major grid disturbance. Also frequency and voltage profile kept as per Indian Electricity Grid Code (IEGC).
- Deployed modern O&M techniques hotline Maintenance, undertaken replacement of porcelain insulators by polymer insulators in pollution prone areas to prevent trippings of lines in foggy winter conditions, and cleaning of insulators through helicopters started since October 2008, first time in India.
- Not a single MW has been bottled up for the want of transmission system.
- Four hierarchical levels of control centres are established, a unique feature in the World with Regional Load Despatch Centres and National Load Despatch Centre (NLDC) at the top.
- To propel POWERGRID in next orbit of growth, Govt. of India has accorded prestigious 'Navratna Status' in May 2008, thereby providing greater autonomy and powers.

3.1.12 Exhibition and Conference on New Technologies—GRIDTECH 2015

GRIDTECH is an international exhibition and conference which provides a platform to manufactures/suppliers/academicians/consultants to showcase their state-of-the-art technologies and products in the field of transmission, distribution, renewable energy integration, smart grid, communications, etc. This biennial international event along with concurrent conference provides a unique opportunity for the power utilities, planners, Policy makers, regulators, manufacturers, research institutions, academicians, consultants, etc., to get exposed to emerging technologies in the above fields for various applications.

GRIDTECH 2015: 5th International Exhibition and Conference
With the above perspective, POWERGRID with the support of the Ministry of Power and in association with CBIP and IEEMA has conducted its 5th International Exhibition and Conference GRIDTECH 2015 from 8 to 10 April 2015 at ITO, Pragati Maidan, New Delhi, on new technologies in transmission, distribution, renewable energy integration, smart grid, communication, etc. The exhibition and the concurrent summit has been an excellent global networking opportunity for exhibitors, visitors and delegates. It has provided an opportunity for all companies to showcase their expertise in various domains of power sector and technology know-how for their awareness, appreciation and benefit overall public sector as well as help them to identify business opportunities in the electricity markets in India and abroad.

The event had been attended by the International/National manufacturers, students, utilities, Policy makers, investors, consultants, academicians, research institutions, etc. It did provide an opportunity to the exhibitor to launch new technology product, for which a well-equipped presentation room had been made available.

Parallel to the GRIDTECH Exhibition, two concurrent conferences for 2 days each had also planned on 8th and 9th April 2015 focusing on state-of-the-art technologies and emerging trends in the field of T&D, renewable energy integration, smart grid, communication, etc.

3.1.13 Awards and Recognitions

The excellent performance of the Company has been recognized and appreciated by the Govt. of India and other prestigious organisations and institutions in form of various awards/accolades in various categories from time to time. The rank, features and other awards and accolades have been mentioned in Table 3.1.

Table 3.1 Rewards and recognition

• Ranked in the PLATTS Top 250 Companies • Features in the Forbes 2000 list
FY 2014–15 • POWERGRID has been conferred 'Star PSU Award' by Business Standard; received by CMD, POWERGRID • 'CA-CFO-Power Sector Award' by Institute of Chartered Accountants of India; received by Director (Finance), POWERGRID • 'The Economic Times Award' by The Economic Times; received by CMD POWERGRID • 'Power Persona of the Year Award' by Central Board of Irrigation and Power (CBIP); received by CMD POWERGRID • 'ICSI National Award for Excellence in Corporate Governance' by Institute of Company Secretaries of India (ICSI); received by Director (Personnel) POWERGRID • 'HR Excellence Award' by Greentech Foundation; received by Director(Personnel), POWERGRID • 'HR Leader Award' by Greentech Foundation; received by Director(Personnel), POWERGRID • 'Organization with Innovative HR Practices Award' by Asia Pacific HRM Congress; received by GM (HRD) • 'JGBS—TOP RANKERS EXCELLENCE AWARD' by Top Rankers • 'STAR PSU CMD OF THE YEAR' by Bureaucracy Today; received by CMD POWERGRID

(continued)

Table 3.1 (continued)

FY 2013–14
• Rated 'Excellent' for its performance as per MoU 2012–13 signed with the Ministry of Power
• Received four (4) Gold Shields and one (1) Silver Shield under 'Comprehensive Award Scheme for recognizing meritorious performance in Power Sector' for Transmission System Availability and Early Completion of Transmission Projects for from Ministry of Power
• Conferred Award for Corporate HR Excellence in Power Sector by 7th ENERTIA Awards 2013, for being India's best place to work for Power Professionals and Power Engineers and
• Conferred 'e-India PSE 2013 Award' for its 'Online Human Resource Development Management System'
• Conferred the Power Line Award 2013 for being the 'Best Performing Transmission Company'
• Bestowed with 'Top Infrastructure Company' under the category 'Power Transmission' by Dun & Bradstreet
• During the year 2013, Chairman & Managing Director, POWERGRID was ranked in '2013-All Asia executive Team—Best CEOs' by 'Institutional Investors', a highly rated North American magazine on international financial market matters
• CMD, POWERGRID has also been conferred with the 'Eminent Electrical Engineer Award' by the Institution of Engineers (India), the largest professional body of Engineers in India
• POWERGRID won the runners-up trophy of National Championship of Business Management Simulations—2013 conducted by All India Management Association (AIMA) and represented India in the Asian Championship
FY 2012–13
• 'MoU excellence award'
• '3rd Rajeev Gandhi Excellence Award 2011' under the 'Best Power Company of the Year'
• 'Fastest Growing Navratna in non-manufacturing category' by Dalal Street Investment Journal
FY 2011–12
• Six projects of POWERGRID bagged the prestigious National Awards for Meritorious Performance in Power Sector instituted by the Government of India for the year 2009–10 and 2010–11
• 'Power Line Awards 2012' in the category 'Best Performing Transmission Company'
FY 2010–11
• 'MoU excellence award'
• 'Emerging telecom infrastructure Provider award'
• Scope award for excellence and outstanding contribution to the Public sector management (2008–09)

(continued)

Table 3.1 (continued)

POWERGRID being a Central Public Sector Enterprise involves mainly in operation which involves national importance and hence, the recognitions received by POWERGRID are mostly at National or International level
Conferred with the prestigious 'MoU Excellence Award' on nine occasions for last 6 years in a row for being top performer in Energy Sector/Syndicate by Department of Public Enterprises, Ministry of Heavy Industries and Public Enterprises and also conferred with Scope Award for Excellence and Outstanding Contribution to the Public Sector Management—Large-scale PSE Category

POWERGRID is poised to excel in the financial year 2014–15 as well in all its five business domains of—operations and maintenance of transmission assets, telecom, grid management, consultancy and project management.

(adapted from www.powergridindia.com, POWERGRID dairy 2014, annual report of POWERGRID for financial year 2013–14 and www.cercind.gov.in).

3.2 Definitions of Terms

Best Practices: Procedures and tasks that have advanced and have proven to be the best.
Collaboration: the course of individuals sharing information and providing contributions to a common goal. Electronic tools perform this as applied to this study.
Community of Practice (*CoP*): a community of practice is a group of workers who gather because of shared skills, ideas and interests of group and it may or may not be part of an official organizational arrangement and the organization does not necessarily mandate the group.
Continuous Variables: a unique number found along a scale but can occur at any point along the scale. Measurement along the scale arrives at continuous variables and they are often fractional because of their nature.
Descriptive Statistics: statistics that display numbers and data about the research target.
Discrete Variable: a unique number determined by reference to a scale. Discrete variables are typically a whole numbers.
Enablers: the tools that enable knowledge workers to manage, capture, store, interpret or recover knowledge.
Epistemology: the philosophical study of knowledge. Epistemology is a study of theories about knowledge and intellect. This study explores knowledge and its nature, sources, limits and validity.
Explicit Knowledge: knowledge that is in storage form or transferable to knowledge workers without adjustment.

Externalization: provides a way to capture and store the knowledge in a central knowledge repository.
Human Capital: the knowledge, skill and talents of the individual workers which comes into existence because of education, experience or exposure. The individual is the owner of this capital.
Implicit Knowledge: knowledge that is not in storable form other than in the mind of a knowledge worker. Implicit knowledge is not easily transferred to an explicit form.
Intellectual Capital: knowledge-based capital in an organization that is of value to the organization. This consists of human capital, social capital and organizational capital.
Internalization: connects employees with the knowledge in the system.
Knowledge Management: management of knowledge. The management can be through logical workflow, physical management or electronic management.
Knowledge Nugget: a piece of knowledge either implicit or explicit. The knowledge object can be of physical or logical form.
Knowledge Worker: an individual who works with the Knowledge Management System. The individual may contribute knowledge, work on knowledge or receive knowledge from the knowledge infrastructure.
Probability Sample: a sample of the universe or whole population where each participant in the sample had a greater than zero chance of selection for the sample.
Random Sample: a sample where each participant in the sample has an equal chance of being chosen as part of the sample.
Rationalism: the theory rationalizing the origin of knowledge and the first modern theory of knowledge.
Simple Sampling: randomly selecting group within the population.
Statistical Sampling: selecting a group of members from the entire population.
Statistical Survey: a method of getting data when a direct approach is not available. Participants provided the data by answering questions on the survey.
Stratified Sampling: simple or systematic sampling of a subgroup with the general population.
Tacit Knowledge: knowledge stored in a knowledge worker's brain. The knowledge is unspoken of and converted into explicit knowledge, spoken, written or explained, before another knowledge worker can gain the knowledge. It is difficult to capture tacit knowledge except through apprenticeship, mentor–mentee shadowing or other such mechanisms.

Reference

POWERGRID (2013), Annual Report—2012–13

Chapter 4
Research Methodology

4.1 Introduction

Research methodology is a science because it is systematic; it requires understanding and correctly using a precise vocabulary, and it requires the competent use of rules. But all aspects of research—from design to implementation to data analysis—can reflect creativity (Loseke 2013). Research design is a conceived plan and structure of investigation to obtain answers to the research question. Research design is defined as the framework or blueprint for conducting the research project. This research is an explanatory research. The research focuses on the impact of several individual and organizational variables that affect the Knowledge Management in POWERGRID. The research methodology adopted is both qualitative and quantitative. The nature of the methodology is exploratory. As the terms suggest exploratory research is conducted because a problem has not been clearly defined as yet and it requires the charting out of the real scope of a problem, exploratory research can be quite informal, relying on secondary research such as reviewing available literature and/or data, or qualitative approaches such as informal discussions with consumers, employees, management or competitors, and more formal approaches through in-depth interviews, Focus Groups, projective methods, case study or pilot study.

Methodology means to be aware of, and be able to handle, different relations which exist between participating moments and process when conducting studies aiming at generating new knowledge (Arbnor and Bjerke 2009). With an aim to generate new knowledge, the inductive reasoning is required in which the premises seek to supply strong evidence for the truth of the conclusion. The philosophical definition of inductive reasoning is much more nuanced than simple progression from particular/individual instances to broader generalizations. Rather, the premises of an inductive logical argument indicate some degree of support (inductive probability) for the conclusion but do not entail it; that is, they suggest truth but do not ensure it. In this manner, there is a possibility of moving from general statement

K.N. Pandey, *Paradigms of Knowledge Management*, Studies in Systems, Decision and Control 60, DOI 10.1007/978-81-322-2785-4_4

to individual instances. At the empirical level, a theory may be developed with inductive reasoning is the logical process of establishing a general proposition of the basis of observation of particular facts. All managers that have ever been seen are human beings; therefore, all managers are human beings (Zikmund 2007). The methodological microcosm of this study has been based on the inductive reasoning and exploratory research for the study of the case.

The research was conducted in two phases and the research methodology will thus be discussed under the two phases. Data in Phase 1 were collected by using a qualitative approach. Qualitative data have been taken from Focus Group opinion, interview, in-depth interview, elite interview and internal observation. In Phase 2, a quantitative approach was followed. Data in this phase were obtained through a structured Questionnaire completed by executives of POWERGRID.

4.2 Research Objectives

(1) To examine whether knowledge is an entity as proffered by epistemological viewpoint or it is an organizational asset which can be managed.
(2) To provide a holistic view of Knowledge Management Implementation to help theory-building effort of Knowledge Management Implementation to emerge.
(3) To identify the causal relation between affecting the Knowledge Management Processes and practices in POWERGRID and with particular emphasis on technological parameters and HR parameters.
(4) To test propositions of Knowledge Management Process and practices with a view to provide a conceptual framework for Knowledge Management Implementation in a Company with many verticals.

4.3 Hypotheses

A hypothesis is something intended for testing. It is a possible explanation, a "might be" for the research. If it is developed correctly, it will point the researcher towards the correct method of data collection. A good hypothesis will have THREE important components. The three parts are as follows: the assumption, the condition and the prediction.

- The assumption is a POSSIBLE EXPLANATION for the problem. This cannot be completed unless the researcher has developed some background knowledge about the subject.
- The condition tells HOW the researcher will conduct the experiment.
- The predictions tell WHAT the researcher thinks will happen.

*H*1: If treated as resource, knowledge is an asset which can be managed.
*H*2: The attitude and behaviour of personnel help Knowledge Management Implementation positively.
*H*3: Culture is the prime mover of Knowledge Management Implementation.
*H*4: Strategy directs KM implementation favourably.
*H*5: Team-based matrix structure affects Knowledge Management Implementation.
*H*6: Well-laid systems enable KM implementation suitably.
*H*7: Predefined processes help Knowledge Management Implementation.
*H*8: Management Support is a prerequisite for Knowledge Management.
*H*9: Relevant training to personnel is a must for Knowledge Management Implementation.
*H*10: Rewards and recognition are motivators for effective Knowledge Management.
*H*11: Information and Communication Technology is an enabler for KM implementation.
*H*12: Tools related with HR are helpful for effective KM.

4.4 Proposed Model

The research model came into existence through reification of the concepts those are prevalent in the literature. Since this is an empirical study, the researcher had seen the intellectual capital of POWERGRID IN THE CONTEXT OF various factors which helped the Knowledge Management Implementation. Since the factors were observed, they required the validation through a model.

4.4.1 Research Model Development

Having completed the review of literature from the point of view of Knowledge Management especially the implementation, it can been seen that culture is all pervasive and it is the most important factor in Knowledge Management and rest of all the factors are having interface with culture. Not only this may 'fit' with the culture, it will be certainly successful. Therefore, the research model depicts culture as an influencing factor of Knowledge Management, overpowering every other factor. Thus, culture has been given the superimposition on every other factor. All the twelve (12) hypotheses which have been emanated from 2.2 to 2.3 beginning with *H*1 to *H*12 have been delineated as the variables which have been mentioned as the factors facilitating the implementation of Knowledge Management in an organization. The proposed model based on hypothesis is given in Fig. 4.1.

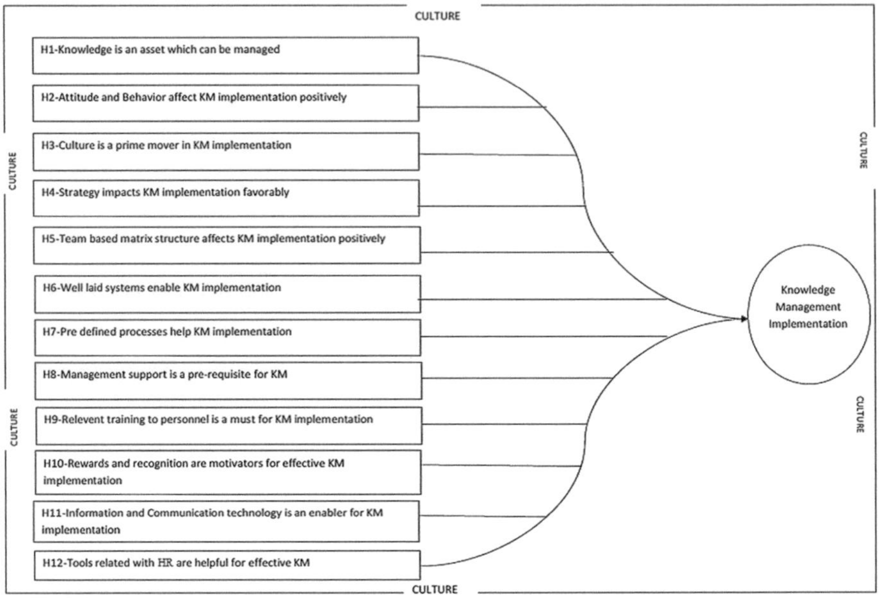

Fig. 4.1 Proposed research model

4.5 Twin Tools for Research

The research was conducted in two phases and the research methodology will thus be discussed under the two phases. Data in Phase 1 were collected by using a qualitative approach. Qualitative data have been taken from Focus Group opinion, interview, in-depth interview, elite interview and internal observation. In Phase 2, a quantitative approach was followed. Data in this phase were obtained through a structured Questionnaire completed by executives of POWERGRID.

4.5.1 Qualitative Tools

The recourse has been taken to use the qualitative tools for this study, and the same has been enumerated as Focus Group, interview, observation, etc.

4.5.1.1 Focus Group

The individuals provide information through Focus Group which is a rich source of primary data. Sekaran and Bougie (2010) have elaborated the Focus Groups in the most lucid style. In their opinion, "Focus Group consist typically of eight to ten members with a moderator leading the discussions for about 2 h on a particular topic, concept, or product. Members are generally chosen on the basis of their expertise in the topic on which information is sought. (….) the focus sessions are aimed at obtaining respondents' impressions, interpretations, and opinions, as the members talk about the event, concept, product, or service. The moderator plays a vital role in steering the discussions in a manner that draws out the information sought, and keeps the members on track. Focus Group discussions are based on a specific topic at a particular location and at a specified time which provides the opportunity for a flexible, free-flowing format for the members. The unstructured and spontaneous responses are expected to reflect the genuine opinions, ideas and feelings of the members about the topic under discussion. Focus Groups are relatively inexpensive and can provide fairly dependable data within a short time frame" (p. 181).

The Focus Group was conducted by the researcher by involving the Nodal Officers of every department and corporate centre, all the regions of the POWERGRID and its Regional Load Dispatch Centres. The selection of and the role played by the moderator is critical. The researcher introduced the topic, observed and took notes and/or taped the discussions. The researcher never became an integral part of the discussions, but merely steered the group persuasively to obtain all the relevant information and helped the group members to get through any impasse that might occur. The researcher also ensured that all members participated in the discussion and that no member dominated the group. It should be noted that although data obtained through these homogeneous group members are less expensive than those obtained through the various other data collection methods and also lend themselves for quick analysis, the content analysis of the data so obtained provides only qualitative and not quantitative information. The Focus Group discussion with Knowledge Management Nodal Officers of POWERGRID representing every department of corporate centre, regional headquarters, various projects and Regional Load Dispatch Centres had been conducted on 27 January 2014. The notes taken in the Focus Group discusses have been placed at Appendix A.

4.5.1.2 Interview

"An interview is a method of data collection that may be described as an interaction involving the interviewer and the interviewee, the purpose of which is to obtain valid and reliable information. Interviews may range from casual conversation of brief questioning to more formal, lengthy interactions, formal interviews are sometimes necessary in research in order to standardize interview topics and general questions. The most important aspect of interviewer's approach concerns conveying the idea that the participant's information is acceptable and valuable" (Marshal and Rossman 1989, p. 82). Interview is an occasion when the researcher puts questions to a person or group of persons. It also entails a session of formal questioning to a group of people. The researcher had conducted two types of interviews: (i) in-depth interview of the committee members who have been entrusted with the work of implementation of Knowledge Management Policy of POWERGRID and (ii) the elite interview of Director (Personnel) of POWERGRID.

4.5.1.3 In-depth Interview

In-depth interviewing is a data collection technique relied on quite extensively by qualitative researchers. It is often described as "a conversation with a purpose" (Kahn and Cannel 1957, p. 149). Once again, this technique can vary depending on the degree the interview is structured beforehand and on the amount of latitude the interviewee is granted responding to the questions. Typically, qualitative in-depth interviews are much more like conversations than formal structured interviews, the researcher explores a few general topics to help uncover the participant's meaning perspective, but otherwise respects how the participant frames and structures the responses. This, in fact, is an assumption fundamental to qualitative research—the participant's perspective on the social phenomenon of interest should unfold as the participant views it, not as the researcher views it.

The analytical categories and instruments of interview have been semi-structured which were designed and carried out in the spirit of qualitative research and have been developed in response to the demands of the material collected. With this in view, in-depth interview of the committee members had been carried out for their usefulness. "Interviews have particular strengths. An interview is a useful way to get large amounts of data quickly. When more than one person is used as an informant, the interview process allows for a wide variety of information and a large number of subjects. It also allows for immediate follow-up questions and, if necessary for clarifications, follow-up interviews may be scheduled at a later date. Combined with observation, interviews allow the researcher to check description against fact" (Marshal and Rossman 1989, p. 82). In-depth interview of Knowledge Management Implementation Committee members of POWERGRID was conducted on 17 January 2014, and the same has been transcribed and appended as Appendix B.

4.5.1.4 Elite Interview

Marshal and Rossman (1989) have differentiated between the interviews into one more category which is known as elite interview. They have depicted that "an interview with an 'elite' person is a specialized case of interviewing that focuses on a particular type of interviewee. Elite individuals are considered to be influential, prominent, and/or well-informed in an organization or community; they are selected for interviews on the basis of their expertise in areas relevant to the research" (p. 105). By taking a cue Shri Ravi Prakash Singh, Director (Personnel), POWERGRID, had been interviewed on 24 January 2014. This elite interviewing had many advantages. Valuable information and futuristic trends have been elicited from him because of his position. He could also provide an overall view of POWERGRID and because of his familiarity with the financial and legal structures of the Company the overall perspective of Knowledge Management could be gleaned. The elite interview has been transcribed and the transcription is available in Appendix C.

4.5.1.5 Internal Observation

Lüders (2004) highlighted the importance of observation by stating that "anyone who wishes to make an empirical investigation of human beings, their everyday practices and life-words has, in principle, two possibilities. One can hold conversations with participants about their actions and collect appropriate documents in the hope of obtaining, in this way, rich information *about* the particular practice in which one is *interested*. Or else one looks for ways or strategies for taking part, for as long as possible, in this everyday practice and becoming familiar with it, so as to be able to observe its everyday performance" (p. 222). Since this researcher is a member of the Knowledge Management Implementation Committee of POWERGRID, he was supposed to be a participant observer and has brought the insight through observations into this research because participant observation is an important root to the sociological description of reality. The role of the participant observer requires both the detachment and the personal involvement in the social milieu of the research. The role of the participant observation has been highlighted by Lüders (2004) as "researchers began to see participant observation in a broader sense as a *flexible, methodologically plural and context related strategy* that could incorporate widely different procedures" (p. 224). Over a period of 3 years, this researcher had been keenly observing various developments in the area of Knowledge Management in POWERGRID.

4.5.2 Quantitative Tools

In order to have overall picture of the study, quantitative tool of survey had been created. It has been opined by Sakeran and Bougie (2010) that researchers adopt 'mixed' methodology comprising of qualitative and quantitative tools; accordingly, the quantitative tool of survey instrument has been used.

4.5.2.1 Development of Survey Instrument

The survey instrument developed to measure the variables was created from various studies conducted earlier in the area of Knowledge Management. The survey instrument has two sections: (i) Personal Information and (ii) Questionnaire. The first section had contained ten questions, and the second section contained 82 questions. While preparing the Questionnaire, the ideas were taken from scholars and a miniscule set of questions were formulated by researcher. The systematic details of the ideas along with their sources are depicted in Table 4.1. The reliability and validity of the instrument had been established through a pilot study.

Table 4.1 Source of questions

Sl. No.	Theme	Question nos.	Source
1	Impediments in Knowledge Management work	3, 4, 5, 6	Questions adapted from Tryon (2012)
2	Knowledge Management strategy	14, 15, 24, 39, 67, 78	Q. Nos. 14, 15, 24 adapted from Chen et al. (2012)
			Q. No. 39, 67, 78 adapted from Shih and Chiang (2005)
3	Culture	1, 2, 7, 8, 9, 10, 28, 36, 70, 71, 73, 74, 77	Q. No. 1, 2, adapted from Tryon (2012)
			Q. No. 7, 8, 9 adopted from Fullan (2011)
			Q. No. 10, 28, 36 adopted from Wood (2005)
			Q. No. 70, 71 adapted from Anderson (2009)
			Q. No. 73, 74, 77 adapted from Swanson (2005)
4	Knowledge as an asset	11, 12, 13, 16, 17	Q. No. 11, 12 adapted from Teece (2001)
			Q. No. 13, 16 adapted from Grant (2001)
			Q. No. 17 adapted from Chen et al. (2012)
5	Reward and recognition	18, 27, 66, 75, 76	Q. No. 18, 27, 66 adapted from Peachey (2006)
			Q. No. 75, 76 adapted from Armstrong (2012)
6	Factors facilitating implementation	19, 20, 21, 22, 23, 33, 34, 35, 37, 38, 49, 53, 57, 62, 72	Q. No. 19, 20, 21 adapted from Pasher and Ronen (2011)
			Q. No. 22, 23 adapted from Upadhyay (2011)
			Q. No. 33, 34, 35, 37, 38 adapted from Almeida (2008)
			Q. No. 49, 53 adapted from Tryon (2012)
			Q. No. 57, 62, 72 adapted from Klein (1998)
7	Structure	25, 26, 40, 41, 44, 46, 47, 56, 58, 65, 80, 81	Q. No. 25, 26, 40 adapted from Brauner and Becker (2006)
			Q. No. 41, 44, 46 adapted from Jones and George (2011)
			Q. No. 47, 56, 58, 68, 80, 81 adapted from Conway and Sligar (2002)

(continued)

Table 4.1 (continued)

Sl. No.	Theme	Question nos.	Source
8	Technology	29, 30, 31, 32, 45, 48, 54, 60, 64, 79,	Q. No. 29, 30, 31 adapted from Man et al. (2008)
			Q. No. 32, 45, 48, 54 adapted from Burg, Berends and Raaji (2008)
			Q. No. 60, 64, 79 adapted from Hasnali (2005)
9	Training	42, 43, 68, 69	Q. No. 42, 43 adapted from Bates (1998)
			Q. No. 68, 69 adapted from Jones and George (2011)
10	Processes	50, 51, 52, 55, 59, 61	Q. No. 50, 51, 52 adapted from Asian Productivity Organisation (2010)
			Q. No. 55, 59, 61 adapted from Anderson (2009)
11	Evaluation	82	Adopted from Ragab and Arisha (2013)
12	Reuse	63	Constructed by researcher

4.5.3 *Pilot Study*

Initially, the survey instrument was prepared by dividing it into eight sections having the headings of General Information, Knowledge Management Environment Assessment, Knowledge Management Strategy, Knowledge Management Tools, Human Resource Management Related Issues, Knowledge Management Practices, Enabling Conditions and Knowledge Management Evaluation in POWERGRID. When the instrument was submitted to the supervisor, she opined that the headings were indicative in nature; hence, these should be dropped. Accordingly, the instrument was recast as one unit with five-point Likert scale, viz. Strongly Disagree, Disagree, Neither Disagree nor Agree, Agree and Strongly Agree. All the questions were put under two heads—Personal Information and Questionnaire having 147 questions. This revised instrument was given to five academicians and ten professionals of POWERGRID from different departments and locations for their comments. It was suggested that the number of questions is very high and it creates unwillingness among them to fill the Questionnaire.

As a result, the constructs of questions were revisited apropos the requirement of the research questions. This examination helped reducing the questions to 82 besides questions of demographic section. The revised instrument was piloted by distributing them to 35 executives (who are knowledge workers). They suggested that response category of Strongly Disagree, Disagree, Neither Agree nor Disagree, Agree and Strongly Agree should be repeated on every page to facilitate smoothness. A suggestion was given that videoconferencing should be added in question number 29 in addition to 'instant messenger' and chat. The participants found that the question number 82 was not clear and it should be rephrased as 'the evaluation is done of work/project that are completed successfully in my organization from KM point of view'. A suggestion was given that the instruction, namely (Please put (✓) mark in the box of your choice) given in Personal Information section, should be repeated again in Questionnaire section.

4.6 Reliability

Reliability analysis allows one to study the properties of measurement scales and the items that comprised the referred scales. The reliability analysis procedure calculates a number of commonly used measures of scale reliability and also provides information about the relationships between individual items in the referred scale. Intraclass correlation coefficients can be used to compute inter-rater reliability estimates. "Reliability applies to a measure when similar results are obtained over time and across situations. Broadly defined, reliability is the degree to which measures are free from error and therefore yield consistent results" (Zikmund 2003, p. 300).

Example Does my Questionnaire measure Knowledge Management in a useful way? Using reliability analysis, we can determine the extent to which the items in our Questionnaire are related to each other, we can get an overall index of the repeatability or internal consistency of the scale as a whole, and we can identify problematic items that should be excluded from the scale.

• *Cronbach's* alpha This model is a model of internal consistency, based on the average inter-item correlation.

Data Data have been coded numerically.

Related procedures To explore the dimensionality of the scale items (to see whether more than one construct is needed to account for the pattern of item scores), the use of factor analysis or multidimensional scaling has been completed and depicted in Tables 4.2 and 4.3.

Table 4.2 Reliability test

Case processing summary			
		N	%
Cases	Valid	49	98.0
	Excluded*	1	2.0
	Total	50	100.0

*Above is equal to data error i.e. 2 % data error exists in data set

Table 4.3 Reliability statistic

Cronbach's alpha	No. of items
0.966	82

4.7 SPSS Output for Cronbach's Alpha

From our example, we can see that Cronbach's alpha is 0.966 as depicted in Table 4.4, which indicates a high level of internal consistency for our scale with this specific sample. This is above the recommended threshold of 0.7 (Hair et al. 2010). Thus, we conclude that the instrument is of good reliability.

Table 4.4 Reliability measurement

Item-total statistics

	Scale mean if item deleted	Scale variance if item deleted	Corrected item-total correlation	Cronbach's alpha if item deleted
VAR00001	295.22	1733.178	0.023	0.966
VAR00002	295.86	1711.375	0.363	0.966
VAR00003	297.02	1743.854	−0.104	0.968
VAR00004	296.67	1758.558	−0.289	0.968
VAR00005	296.98	1742.229	−0.088	0.967
VAR00006	297.06	1745.559	−0.130	0.967
VAR00007	296.22	1687.136	0.569	0.966
VAR00008	296.18	1696.778	0.530	0.966
VAR00009	296.33	1692.808	0.573	0.966
VAR00010	295.82	1710.736	0.348	0.966
VAR00011	295.73	1698.366	0.457	0.966
VAR00012	296.10	1712.510	0.302	0.966
VAR00013	295.59	1704.705	0.524	0.966
VAR00014	295.96	1684.957	0.636	0.965
VAR00015	296.18	1678.195	0.686	0.965
VAR00016	296.14	1687.125	0.608	0.966
VAR00017	296.31	1685.467	0.560	0.966
VAR00018	296.59	1695.955	0.436	0.966
VAR00019	296.39	1699.909	0.518	0.966
VAR00020	296.63	1678.237	0.653	0.965
VAR00021	296.22	1686.761	0.519	0.966
VAR00022	295.78	1719.719	0.251	0.966
VAR00023	295.96	1698.248	0.531	0.966
VAR00024	296.08	1689.785	0.584	0.966
VAR00025	296.14	1694.083	0.516	0.966
VAR00026	296.18	1687.445	0.629	0.966
VAR00027	296.76	1695.689	0.509	0.966
VAR00028	296.24	1674.480	0.669	0.965
VAR00029	297.00	1685.167	0.458	0.966

(continued)

Table 4.4 (continued)

Item-total statistics				
	Scale mean if item deleted	Scale variance if item deleted	Corrected item-total correlation	Cronbach's alpha if item deleted
VAR00030	296.67	1693.349	0.426	0.966
VAR00031	296.57	1678.458	0.683	0.965
VAR00032	297.12	1677.985	0.578	0.966
VAR00033	295.90	1700.135	0.451	0.966
VAR00034	295.78	1709.636	0.367	0.966
VAR00035	297.14	1704.667	0.285	0.966
VAR00036	296.10	1690.885	0.538	0.966
VAR00037	296.45	1692.961	0.462	0.966
VAR00038	296.92	1680.327	0.567	0.966
VAR00039	296.49	1670.797	0.724	0.965
VAR00040	296.27	1698.616	0.480	0.966
VAR00041	296.57	1677.250	0.669	0.965
VAR00042	295.92	1690.910	0.437	0.966
VAR00043	296.04	1676.915	0.622	0.965
VAR00044	296.33	1687.766	0.568	0.966
VAR00045	296.41	1670.580	0.651	0.965
VAR00046	296.76	1675.189	0.662	0.965
VAR00047	295.84	1692.806	0.517	0.966
VAR00048	295.73	1690.282	0.577	0.966
VAR00049	295.94	1685.267	0.638	0.965
VAR00050	296.10	1689.094	0.589	0.966
VAR00051	295.71	1708.750	0.490	0.966
VAR00052	295.96	1694.332	0.512	0.966
VAR00053	295.82	1711.778	0.357	0.966
VAR00054	295.55	1695.711	0.646	0.966
VAR00055	295.61	1711.701	0.354	0.966
VAR00056	295.69	1710.092	0.367	0.966
VAR00057	295.41	1713.455	0.459	0.966
VAR00058	296.39	1669.784	0.692	0.965
VAR00059	296.67	1681.016	0.550	0.966
VAR00060	296.33	1702.308	0.415	0.966
VAR00061	295.86	1693.542	0.609	0.966
VAR00062	295.61	1702.534	0.483	0.966
VAR00063	295.57	1724.875	0.190	0.966
VAR00064	296.43	1673.458	0.676	0.965
VAR00065	296.45	1671.919	0.648	0.965
VAR00066	296.45	1664.253	0.687	0.965

(continued)

Table 4.4 (continued)

Item-total statistics

	Scale mean if item deleted	Scale variance if item deleted	Corrected item-total correlation	Cronbach's alpha if item deleted
VAR00067	296.88	1678.735	0.594	0.966
VAR00068	296.24	1679.522	0.589	0.966
VAR00069	295.84	1699.098	0.553	0.966
VAR00070	295.88	1709.610	0.347	0.966
VAR00071	296.08	1671.660	0.698	0.965
VAR00072	295.82	1685.528	0.670	0.965
VAR00073	295.94	1691.059	0.576	0.966
VAR00074	295.90	1673.427	0.759	0.965
VAR00075	296.76	1681.480	0.540	0.966
VAR00076	296.65	1679.815	0.612	0.966
VAR00077	296.47	1677.546	0.656	0.965
VAR00078	296.29	1659.208	0.719	0.965
VAR00079	295.65	1704.273	0.421	0.966
VAR00080	296.65	1667.898	0.700	0.965
VAR00081	296.61	1664.951	0.703	0.965
VAR00082	296.51	1668.338	0.702	0.965

Reliability Algorithms

There are two methods: Method 1 and Method 2 prevalent in statistics when it entails two different computing methods, depending upon the model specification and options. Method 1 does not involve computing a covariance matrix. It is faster than Method 2, and it is used for large problems which requires much less work-space. Method 2 requires computing a covariance matrix of the variables. It is slower than Method 1 and requires more space. However, it can process all models, statistics and options. The two methods differ in one other important respect. Method 1 will continue processing a scale containing variables with zero variance and leave them in the scale. Method 2 will delete variables with zero variance and continue processing if at least two variables remain in the scale. If item deletion is required, Method 2 can be selected by requesting the covariance method. Further analysis will be done by using the above-mentioned two methods in order to get improve the results.

At this juncture, it is to mention that various ideas were called out from different researchers and authors on various themes in order to prepare the questions in the instrument.

This column presents the value that Cronbach's alpha value would change if that particular item was deleted from the scale. We can see that removal of any question, except question "The knowledge once used in my department is reused in similar situations again", would result in a lower Cronbach's alpha. Therefore, we would not want to remove these questions. Removal of question "The knowledge once used in my department is reused in similar situations again" would lead to a small improvement in Cronbach's alpha, and we can also see that the "corrected item-total correlation" value was low (0.190) for this item. Therefore, we may delete this question because it has been contributed by this researcher and therefore does not purport to rest of the questions which have been adopted from previous studies of scholars. We are not considering here the deletion of questions 1–6 because they are directly related with the objective of the study and the same are required for testing of the hypothesis. It is believed that if sample size will be increased, then it would result in improvement in Cronbach's alpha.

Cronbach's alpha simply provides us with an overall reliability coefficient for a set of variables (e.g. questions). If our questions reflect different underlying personal qualities (or other dimensions), e.g. reward, recognition and training, Cronbach's alpha will not be able to distinguish between these. In order to do this and then check their reliability (using Cronbach's alpha), we first need to run a test such as a principal components analysis (PCA).

4.7.1 Validity

"Validity is the ability of a measure to be measured what is supposed to measure. If it does not measure what it is designated to measure, there will be problems" (Zikmund 2003, p. 302). Bharti and Chaudhary (2004) assert that content validity is an essential prerequisite prior to further statistical analyses. Content validity refers to the level of representativeness of construct to its facets; hence to ensure content validity, all the constructs were adopted from previous studies (Lee 2013).

4.7.2 Final Survey Instrument

After the pilot test of the Questionnaire, reliability and validity were checked and survey instrument was accordingly revised. The final Questionnaire for survey of Knowledge Management in POWERGRID has been appended as Appendix D.

4.7.3 Sampling

On the basis of research development, population size of executive class of employees was adhered to as number of executives in POWERGRID is 3990 because the executive class of employees are the 'knowledge workers'. Structured close-ended Questionnaire had been developed and distributed randomly among 1000 executives covering all the 5 business verticals and its Enablers of

POWERGRID. Out of these 1000 executives, only 525 responses were received. Out of these 525 responses, 414 responses were found valid.

4.8 Assumptions of the Study

Observations should be independent, and errors should be uncorrelated between items. Data are based on five-point Likert scale, so its property signifies that the data are nonparametric (Kunzon et al. 1996). Scale should be additive, so that items are assumed to be linear. It is assumed that time taken in the survey is zero and therefore has no impact on responses. The executives posted at various locations in India and abroad have responded to questions without having bias arising from the place of posting.

4.8.1 Normality

Maximum likelihood estimation is robust to no normality. Parameter estimates are efficient, and structural equations (SEs) are lightly biased. If bootstrap has been done, then there is no normality requirement at all (SEs are asymptotic bootstraps) unless you select the "parametric bootstrap" (João 2014).

The ML (maximum likelihood) estimator is considered relatively robust to violations of normality assumptions (Diamantopoulos et al. 2000). The aim of maximum likelihood estimation is to find the parameter value(s) that makes the observed data most likely. This is because the likelihood of the parameters given as the data is defined to be equal to the probability of the data given by the parameters. (Technically, they are proportional to each other, but this does not affect the principle.)

Multivariate normality (mvm) is sufficient but not necessary for normal theory maximum likelihood for SEM. Two counter-examples are given as follows: Two groups each normally distributed, but their means differ. Here, the data from the full sample are non-normal (a mixture of two multivariate normal). Ordinal data are analysed by full information to maximum likelihood. Here, the likelihood is computed by integrating the multivariate normal distribution, so it is a normal theory with maximum likelihood but with, e.g., a threshold model that separates the ordinal categories. Yet the data themselves are decidedly non-normal (Michael 2014).

4.8.2 Software Used

SPSS V.20–V.21 had been used for reliability tests and analytical statistics including factor analysis (Berkman and Reise 2012). This software had been used as tool to analyse quantitative data. So far as qualitative data analysis is concerned, NVivo V.10 software had been used and documents were imported from word and coded on screen (Grbich 2013).

References

Almeida, P.: Localization of knowledge and the mobility of engineers in regional networks. In: Kogut, B. (ed.) Knowledge, Options and Institutions. Oxford University Press, Britain (2008)

Anderson, K.K.: Organizational capabilities as predictors of effective knowledge management: an empirical examination, Ph.D. Thesis of H. Wayne Huizenga School of Business and Entrepreneurship, Nova Southern University (2009)

Arbnor, I., Bjerke, B.: Methodology for Creating Business Knowledge, 3rd edn. Sage Publication India Pvt. Ltd., New Delhi (2009)

Armstrong, M.: Armstrong's Handbook of Human Resource Management Practice, 12th edn. Kogan Page Ltd., London (2012)

Asian Productivity Organisation: Knowledge Management Tools and Techniques Manual. Tokyo, Japan (2010)

Bates, W.R.J.: Organizational Learning and Knowledge Technologies in a Dynamic Environment. Kluwer Academic Publishers, Boston (1998)

Berkman, E.T., Reise, S.P.: A Conceptual Guide to Statistics Using SPSS. Sage, L.A. (2012)

Bharti, P., Chaudhary, A.: An empirical factors investigation of decision making-making satisfaction in web based decision support systems. Decis. Support Syst. **37**(2), 187–197 (2004)

Braumer, E., Becker, A.: Knowledge management in interaction: transactive knowledge systems and the management of knowledge. In: Renzl, B., Matzler, K., Hinterhulber, H. (eds.) The Future of Knowledge Management. Palgrave Macmillian, New York (2006)

Burg, E.V., Berends, H., Raaji, E.V.: Organizing knowledge sharing in networks: the theory. In: Man, A.P.D., Matzler, K., Hinterhulber, H. (eds.) Knowledge Management and Innovation in Networks. Edward Elgar, Cheltenham, UK (2008)

Chen, Y.Y., Yeh, S.P., Huang, H.L.: Does knowledge management "fit" matter to business performance? J. Knowl. Manage. **16**(5), 671–687 (2012)

Conway, C., Sligar, C.: Knowledge Management: Unlocking <K_n_o_w_l_e_d_g_e> Assets solutions from Microsoft. Prentice-Hall of India Pvt. Ltd., New Delhi (2002)

Diamantopoulos, A., Siguaw, J., Siguaw, J.A.: Introducing LISREL: A Guide for the Uninitiated. Sage Publications, London (2000)

Fullan, M.: Change Leader: Learning to Do What Matters Most. Jossey-Bass, A Wiley Imprint, San Franscisco, CA (2011)

Grant, R.M.: Knowledge and organization. In: Nonaka, I., Teece, D.J. (eds.) Management Industrial Knowledge: Creation, Transfer and Utilization. SAGE Publications, London (2001)

Grbich, C.: Qualitative Data Analysis: An Introduction. SAGE Publications Ltd., New Delhi (2013)

Hair, J.F., Black, W.C., Balin, B.J., Anderson, R.E.: Multivariate data analysis. Maxwell Macmillan International Editions, New York (2010)

Hasnali, F.: Building a knowledge-sharing network: plan, design, execute…. reap?. In: Rao, Manmohan (ed.) Knowledge Management Tools and Techniques. Elsevier, New York (2005)

João, M.: Google scholar accessed on 16th November by using structural equation modeling for ordinal data. Board of Directors of IAVE, I.P. and Associate Professor of Statistics at ISPA-IU (2014)

Jones, G.R., George, J.M.: Contemporary Management. McGraw-Hill, Irwin (2011)

Kahn, R., Cannel, C.: The Dynamics of Interviewing. John, Wiley, New York (1957)

Klein, D.A.: The Strategic Management of Intellectual Capital: An Introduction. Butterworth-Heinemann, Boston (1998)
Kuzon, W.M., Jr., Urbanchek, M.G., McCabe, S.: The seven deadly sins of statistical analyses (1996)
Lee, V.-H., et al.: Knowledge management: a key determinant in advancing technological innovation. J. Knowl. Manage. **17**(6), 848–872 (2013)
Loseke, D.R.: Methodological Thinking: Basic Principles of Social Research Design. Sage Publications Incorporated, New Delhi (2013)
Lüders, C.: Field observations and ethnography. In: Flick, U., Kardorff, E.V., Steinke, I. (eds.) A Companion to Qualitative Research. Sage Publications Ltd., London (2004)
Man, A.P.D., Berends, H., Lammers, I., Raaij, E.V., Weele, A.V.: Knowledge and innovation in networks: a conceptual framework. In: Man, A.P.D. (eds.) Knowledge Management and Innovation in Networks. Edward Elgar, Cheltenham, UK (2008)
Marshall, C., Rossman, G.B.: Designing Qualitative Research. Sage Publications, Thousand Oaks, London, New Delhi (1989)
Michael, N.: Google scholar accessed on 16th November by using structural equation modeling for ordinal data Professor at Virginia Commonwealth University) (2014)
Pasher, E., Ronnen, T.: The Complete Guide to Knowledge Management: A Strategic Plan to Leverage your company's Intellectual Capital. John Wiley & Sons Inc., Hoboken, New Jersey (2011). 188
Peachey, T.A.: An examination of the effects of cultural, climatic, structural and technological factors on knowledge management effectiveness, Ph.D. Thesis; Asburn, Alabama (2006)
Ragab, M.A.F., Arisha, A.: Knowledge management and measurement: a critical review. J. Knowl. Manage. **17**(6), 873–901 (2013)
Sekaran, U., Bougie, R.: Research Methods for Business—A Skill Building Approach, 5th edn. Wiley India, New Delhi (2010)
Shin, H.A., Chiang, Y.H.: Strategy alignment between HRM, KM and corporate development. Int. J. Manpower **26**(6), 582–606 (2005)
Swanson, T.C.: Knowledge management: an analysis of knowledge integrator effectiveness in managing organizational knowledge, Ph.D. Thesis of Capella University (2005)
Teece, J.D.: Strategies for management of knowledge asset: the role of firm structure and industrial context. In: Nonaka, I., Teece, D.J. (eds.) Management of Industrial Knowledge: Creation, Transfer and Utilization. SAGE Publications, London (2001)
Tryon Jr, C.A.: Managing Organisational Knowledge 3rd Generation Knowledge Management … And Beyond. CRC Press Taylor & Francis Group, London (2012)
Upadhyay, G.: HR Interventions in the Age of Competitiveness: Analysis of HR-Based Change Interventions Introduced in a Large Manufacturing Organizations. Reem Publications Pvt. Ltd., New Delhi (2011)
Wood, C.: An empirical Examination of factors influencing work—unit knowledge management effectiveness in organizations, Ph.D. dissertation. University of Texas at Arlington (2005)
Zikmund, W.G.: Business Research Methods. Thomson, South-weateern (2003)
Zikmund, W.G.: Business Research Methods. Third Indian Reprint, Thomson, South Western, Delhi (2007)

Chapter 5
Data Collection, Analysis and Interpretation

5.1 Introduction

Organizations have installed and configured Enablers of KM systems, but few have set up a full KM infrastructure. This study has been carried out by using qualitative and quantitative both methods. The data had been collected about the status as how Knowledge Management has been in practice in POWERGRID. First, the qualitative data analysis has been delineated followed by quantitative analysis of data collected through survey responses. In this context, the journey of Knowledge Management in POWERGRID has been traced.

5.2 The Odyssey of Knowledge Management in POWERGRID

This section will provide a bird's-eye view of the journey of Knowledge Management in POWERGRID from envisioning to evaluation. Since public sector companies in India are still continuing with their bureaucratic structure, the management of POWERGRID started by establishing a Knowledge Management Cell followed by formulation of Knowledge Management Policy and creation of Knowledge Bank. At this juncture, this case study should analyse various tools, drivers and other Enablers such as information and communication technology in detail.

5.2.1 Knowledge Management Cell

POWERGRID started its journey in a typical public sector manner by creating a Knowledge Management Cell in April 2009 with a view to manage the knowledge

K.N. Pandey, *Paradigms of Knowledge Management*, Studies in Systems, Decision and Control 60, DOI 10.1007/978-81-322-2785-4_5

in Company. There were two factors—attrition and superannuation—which necessitated the Knowledge Management in this Company. Any of these factors may lead to the loss of requisite knowledge of Company which may result in incapacitation of any function. It was therefore, important to develop a mechanism so that knowledge could be managed in Company and the onslaught of loss of knowledge due to these two factors could be mitigated. POWERGRID has witnessed the attrition at an accelerated pace in the young executives and superannuation of senior executives which had been hampering its knowledge flow in a big way. Therefore, they did take recourse to Knowledge Management on an emergent basis in order to keep their lead in the areas of transmission of bulk power, grid management and telecom to keep their consultancy project afloat. Knowledge Management Cell was established by appointing one chief manager (HR), one manager (HR), an IT engineer, two electrical engineers and two civil engineers.

The recently established Knowledge Management Cell formed a committee comprising of executive director (HR), executive director (Engineering) and deputy general manager (IT) to formulate a Knowledge Management Policy besides preparing a Glossary of Knowledge Management so that the personnel who were to prepare the Policy could get familiarized with the terms used in the field of Knowledge Management. The committee visited various organizations such as Bharat Heavy Electricals Ltd., Indian Oil, NTPC, TCS and TISCO. The committee interacted with the experts managing the knowledge in their respective companies and also collected their respective Knowledge Management Policies. After deliberation in various meetings, Company could formulate and publish its Knowledge Management Policy in August 2010 (POWERGRID 2010).

5.2.2 *Knowledge Management Policy*

The Knowledge Management Policy comprised of Knowledge Management Definition; Objectives; Rationale for Knowledge Management; Knowledge Management Strategy; Virtual KM Organization; KM Procedure; Knowledge Bank; KM Practices for People; Incentives and Recognition. In addition to these, it also provided Scope/Application; Policy Framework; Role of Knowledge Management Cell, i.e. Role of Individual Staff, Role of Nodal Officers, Role of Heads of the Departments; Compliance; Interpretation and Revision of the Policy. The mission and objectives of the Policy are as follows:

Mission: 'Acquire, Create, Analyse, Share and Use knowledge strategically for enabling POWERGRID to become a pioneer Company in its business domains through Knowledge Management Processes and strategies tailored to facilitate and motivate people towards knowledge sharing' (POWERGRID 2010, p. 7).

Vision: POWERGRID will be knowledge provider to further the capabilities of its personnel for creating knowledge sharing culture to increase business in all its verticals by reducing time and cost.

Objectives: The objectives of the Knowledge Management in POWERGRID entails—'Harnessing potential of Information and Communication Technology; Increasing the organizational process efficiency and accuracy for maximizing the returns; Enriching knowledge contents for establishing and creating new ways and means in business process of organization; Facilitating and managing organizational innovation and learning; Leveraging the expertise of people across the organization; Increasing the knowledge exchange among employees; Managing intellectual capital and intellectual assets in the work force; Adopting new Knowledge Management tools, techniques and technologies to keep the KM system in POWERGRID intact, robust and going; Increasing employee satisfaction by facilitating employees learning' (POWERGRID 2010, pp. 9–11).

5.2.3 *Nodal Officers for Knowledge Management*

While formulating the Policy, the Company was simultaneously nominating nodal officers for implementation of Knowledge Management. Accordingly, in August 2009, Nodal Officers were nominated in every department, region and Regional Load Dispatch Centres. While nominating the Nodal Officers, it was envisaged that Knowledge Management is an onerous task and it requires the formation of virtual organization to implement the Knowledge Management across the length and breadth of the organization. As a prelude to this, a Nodal Officer was required to be earmarked from every region/department who should be computer savvy and evangelistic in nature besides having in-depth knowledge of his department/region/RLDC to coordinate with Knowledge Management Cell in implementation of Knowledge Management programme of Company. They would also help in identifying the persons who would capture the knowledge and information related to the respective department that was required to be managed for reuse in future. Once the virtual organization of KM came into being, Nodal Officers were providing data and information regarding the departmental experts who would edit the contributed pieces of information and knowledge by personnel of his department.

5.2.4 *Knowledge Management Awareness Programmes*

The Company was aware that a ripple effect should be created in the Company. By keeping this in view, Knowledge Management Awareness Programmes were simultaneously organized throughout the Company by a fiat. In order to disseminate the contents of Knowledge Management Policy and to create the awareness of Knowledge Management Implementation, a one-day 'Awareness Programme on Knowledge Management' was organized across the Company. With this in view, every region/RLDC had completed conducting the awareness programme on 31 March 2011.

5.2.5 Pilot Implementation of Knowledge Management Policy

Knowledge Management Cell of POWERGRID implemented Knowledge Management Policy in Company as a Pilot Project in its southern region comprising of Tamil Nadu, Karnataka and Kerala. The pilot was completed by September 2011, and in the implementation report, it was found to be effective in the areas of autotransformer and reactor; CBs and isolators; protection systems; AC system, FF system, battery and chargers; transmission lines (construction, O&M and HLM); substation (construction and O&M), contract and material; F&A, audit, cost estimation; and HR, IE, CSR. However, when the Policy was being piloted, it was found that the enabling functions of finance were not in consonance with the rest of the functions of the region. Accordingly, the Policy was revised to address the concern of finance department in December 2011.

5.2.6 Knowledge Management Technology

While POWERGRID was on a spree of undertaking various activities for Knowledge Management, it was found that the existing information and communication technology was not sufficient so it undertook the exercise to:

- Review of existing technology available in POWERGRID,
- Appraisal of existing system and technology in peer organizations who have effectively implemented Knowledge Management system,
- Identification of industry leader in the relevant field by gleaning analytical reports and market survey,
- Identification of alternative options for finally selecting the Knowledge Management Technology.

As a result of this introspection, POWERGRID acquired requisite technology to ensure the following:

- A homepage to assert existence of Communities of Practice/Team Sites in POWERGRID and describe domain and activities.
- Conversation space for online discussions.
- Repository for the documents, including research reports, best practices and standards.
- Team sites for explicit knowledge sharing as per knowledge map prepared by POWERGRID.
- A good cross-site search engine to find things in the knowledge base.
- A directory of membership with some information about members' areas of expertise in the domain.

- In some cases, a shared workspace for synchronous electronic collaboration or to enhance teleconference with visuals.
- BI tools for creating visual dashboards from data in portal or in custom application.
- Workflow for content approval and other business process as may be required.
- Community management tools, mostly for the coordinator but sometimes also for the community at large. These might include the ability to know who is participating actively, which documents are downloaded, how much traffic there is and which documents need updating.

POWERGRID also found that Enterprise Resource Planning Technology from SAP should be taken and implemented across the length and breadth of the Company. As a part of ERP, Knowledge Management Technology was also upgraded in the year 2010.

5.2.7 *Knowledge Bank*

Knowledge Bank was established in January 2012 to enable the Knowledge Management Processes of Creation and Acquisition; Collation; Storage and Use; Dissemination; Sharing; Reuse; Capitalization to come to fruition. The creation of knowledge in POWERGRID has been taking place through Technology Development Department. The creation of tacit knowledge has been taking place in the minds of personnel of POWERGRID which get posted in the Knowledge Management Portal known as Knowledge Bank. The Company has been using Knowledge Management Technology through Information and Communication Technology Tools. POWERGRID has also been acquiring knowledge through networking with suppliers, stakeholders, vendors, customers, regulators and power traders. So far as the collation of knowledge nuggets is concerned, POWERGRID is undertaking this exercise by maintaining and updating its Knowledge Bank on Power Portal. All the employees of POWERGRID were given excess to take the knowledge nuggets from Knowledge Bank. Knowledge Bank was envisaged to ensure the following:

(a) 'People Search—Expert pages (A list of Domain Leaders), Project owners, Vendor information, Customer information.
(b) Training material—FAQs, Presentations, Audio/Video files, Orientation courses, reading material, Self-Study and evaluation courses for a specific role.
(c) Literature—Books/Manuals, White papers/Research articles.
(d) Discussion Forums/Blogs.
(e) Lessons Learnt—Known issues and workarounds for Project Issues as well as Maintenance Issues, experiential knowledge.
(f) External resources—Link of external websites for reference, Book Reviews, External Environment Inputs like Regulations etc., Innovations/Latest Technology.

(g) Corporate Information—presentations, Key Performance Indicators like Budget/Availability/Revenue/Telecom Revenue etc., Circular Repository, CSR information.
(h) Project information—Progress Reports, MIS Reports as given in departmental ISO procedures, Measures and Metrics.
(i) Technical Drawings, Specifications and blue prints for various component, equipments and Systems.
(j) Technical Reports—Test Results, Outage Reports, Failure Analysis Reports, Assessment reports.
(k) Records of meeting—PRM/ORM Agenda and Minutes, Vendors meet Agenda/Minutes, Minutes of PNBC Meetings, Minutes of HOP meetings.
(l) Competitor's Information.
(m) Buzzwords.
(n) Glossary of frequently used terms and abbreviations' (POWERGRID 2010, pp. 23–25).

Besides taking care of all the Knowledge Management processes, it has also given a foundation for dissemination of knowledge and functions of various Communities of Practice.

5.2.8 Task Force for Implementation of Knowledge Management

Once the Knowledge Management Policy was implemented in the Company, it was felt that there should be a Task Force for implementation of Knowledge Management. This Task Force was constituted by taking one executive from every strategically important department comprising of Information Technology, Corporate Planning, Engineering, Finance, Operations Servicers, System Operation and Human Resources. The name of the executives of Task Force for Implementation of Knowledge Management is given in Appendix E.

5.2.9 Knowledge Audit and Publication of Knowledge Maps and Matrices

POWERGRID conducted an audit of knowledge and after that it published its Knowledge Maps and Matrices (POWERGRID 2011a). This publication became the important tool in the hands of knowledge workers so far as the explicit knowledge is concerned. POWERGRID has claimed that 'the knowledge Maps and Matrices of POWERGRID has been organized under the head of Strategy, Knowledge Requirements, Knowledge Gap, Knowledge Assets and Knowledge Domains of the Department. The Knowledge Matrix has been organized in the columns of What Knowledge is Needed, Who Has It, Who Needs It, Where Is It, Is It Tacit or Explicit, Is It Routine or Non-routine and What Issues does it address' (POWERGRID 2011a, p. iii). Knowledge Maps and Matrices can be used for the dissemination of knowledge and to glean the information for inter-departmental meetings for sharing the knowledge for developing new functions for benefit of POWERGRID. Knowledge Maps and Matrices can be used for acquiring, analysing, sharing, using and reusing of knowledge for innovative purposes. Knowledge Maps and Matrices are an explicit repository of knowledge, information, conceptual resources for developing new business opportunities, help facing the challenges and mitigating difficulties of POWERGRID.

5.2.10 Corporate Yellow Pages (Subject-Matter Experts)

Knowledge Management entails that the tacit knowledge is made available to those who need it. Tacit knowledge includes insights and experiences of the knowers which can the made available to those who require it by providing the information about subject-matter experts in an organization. Knowledge is recognized as a key source of competitive advantage in the business world. Traditional Knowledge Management approaches attempt to capture existing knowledge within formal systems, such as repositories and enabling databases. Yet systematically addressing the kind of dynamic 'knowing' that makes a difference in practice requires the participation of people who are fully engaged in the process of creating, refining, communicating and using knowledge. With the arrival of knowledge economy companies have to manage their knowledge assets, share, use, reuse, capitalize them and thus remain ahead of the competition. With this in view, POWERGRID intended to become a knowledge-driven Company to keep pace with the requirement of knowledge economy by preparing the knowledge assets to leverage it in forthcoming projects and functions. Preparation of Corporate Yellow Pages was a step in this direction. This assumes human resource-oriented approach and views information as a tool to support the creative genius in the seemingly unlimited potential of an individual's knowledge.

One of the most common problems in Knowledge Management is how to find expert help when there arises a problem. No matter how trained and experienced employees are, they cannot know everything. Most of the personnel come across new problems and questions and the Company has to help them find answers. Composing Yellow Pages for the organization is a relatively easy task from an information technology point of view, but a rather challenging task from a cultural point of view. The task involves addressing the following issues:

- How to identify and choose the experts and
- How to ensure that the experts are available and willing to help in spite of their own busy schedule.

In order to address the above concerns, it was imperative for POWERGRID to prepare Corporate Yellow Pages (subject-matter experts) and maintain this repository so that it could be uploaded on Knowledge Bank. With this in view, POWERGRID prepared Corporate Yellow Pages on the lines of business verticals —Operation and Maintenance of Transmission Assets; Telecom; Grid Management; Project Management; Consultancy and Enablers—Vigilance; Common collaboratory of Core Expertise; Human Resource; Finance; Information Technology; and Miscellaneous. It formulated that the purpose of corporate Yellow Pages is to facilitate the dissemination of tacit knowledge between the person who needs it and the knower. It was organized in the following format (Fig. 5.1).

Having prepared the Corporate Yellow Pages, it was electronically published on 30 January 2012 besides uploading it on Knowledge Bank and HRD Website on Power Portal.

5.2.11 *Initiation of Communities of Practice*

POWERGRID initiated development of various Communities of Practice (CoP) in the year 2011. The management of POWERGRID was intending to foster willingness for creating communities that have both a sense of purpose and shared values and rules of engagement that are designed to encourage collaboration, discovery-driven learning and integrative decision-making. Fostering ability

Emp. No.	Names (Sh.)	Designation	E-Mail	Mobile

Professional Area :

Areas of Specialization :

Type of Expertise :

Brief Description of experience/exposure/problems faced/ solutions offered/ tools and drivers used :

-
-
-
-

Unique Point (if any) :

-
-

Fig. 5.1 Template of yellow pages

requires developing these organizational capabilities: creative abrasion, creating agility and creation resolution. To build willingness, the management created communities which share a sense of purpose and rules of engagement. The tenets were affirming the individuals and the groups, supporting and confronting, fostering experimentation and learning. These were having an eye on promoting improvisation and structure; showing patience and urgency; encouraging bottom-up initiative and intervening top-down. Various Communities of Practice on Operation and Maintenance, Telecom, Grid Management, Project Management and Consultancy along the business verticals of POWERGRID were formed having a subject-matter expert as an Anchor, who was also an influential person. In addition to these, three more Communities of Practice were created on Finance, Human Resources and Information Technology. The detailed description of these 8 Communities is given in Appendix F.

Once the Communities of Practice were created, the same were given vide publicity in Company. POWERGRID circulated that Knowledge Management Policy had been formulated in the Company with a view to create an environment of free and open knowledge sharing. In this direction, various Communities of Practice for all the business verticals had been cultivated. It is operational under Knowledge Bank of Power Portal under the subhead Knowledge Management Site. It can be accessed at http://sps.powergridindia.com/KNOWN%20ERROR%20DATABASE/Pages/dafault.aspx. This platform can be utilized for the purpose of sharing of knowledge through contributing or using the knowledge and information through papers, articles, etc., which can be used towards success of the various functions in a meaningful manner towards increasing the human capital of the Company for gainful purposes.

In order to give a fillip and buttress, the Communities of Practice POWERGRID also circulated an action plan. It purported that the CoPs would aim at:

- Support effective integration/implementation of Knowledge Management.
- Support a specific topic or development challenge, e.g. aid effectiveness, by stimulating an active sharing of knowledge and problem solving.

5.2.11.1 CoPs Responsibilities

(1) Anchors: To ensure continuity, CoPs should have a coordinator and a deputy coordinator or two co-coordinators. The coordinator(s) of a CoP has overall responsibility for ensuring that it achieves its objectives and work plan effectively and efficiently and for preparing the annual report for POWERGRID. Coordinators and their managers should agree on the level of commitment that is required. In POWERGRID case, they have christened them as Anchors.

(2) The IT administrator reports to the coordinator (Anchor) and provides administrative support in pursuit of the CoP's purpose and objectives.

(3) Core members contribute to the effective and efficient achievement of the CoP's purpose and objectives and its work plan and may take the lead on specific activities. Core members each report on the CoP work plan and activities to their group and each communicates the interests and concerns of their group of the CoP.
(4) CoP members contribute to the achievement of the CoPs purpose and objectives and the development and implementation of its work plan, including outreach activities.

5.2.11.2 Commitment from CoPs

CoPs only work effectively if members not only have enthusiasm and interest, but also have sufficient time available to participate in CoP meetings and activities and receive recognition for participating through their work plan and the performance management process. It entailed in POWERGRID that:

- CoP members should work with their line managers to agree on the level of commitment that can be made. Being a core members serving as a representative of a group and engaging in CoP activities requires a substantial commitment of time. Other members may take a lesser role or have limited engagement. However, it is expected that all member will take on the role of being a 'champion' for the CoP.
- CoP members should record their roles, responsibilities and time commitments in individual performance agreements and monitor these. It is important that the HODs facilitate and support staff participation on CoPs and recognize staff contributions in performance management processes. With the above in view, following actions are to be taken by the Anchors of the Communities of Practice (CoPs).

5.2.11.3 Actions to Be Taken by CoPs

A CoP prepares an annual work plan. While the specific content will vary, the work plan should reflect the overall objectives of a CoP as set out in the term of reference (e.g. putting Policy into good practice; promoting good practice, developing skills, and promoting learning; and sharing information through outreach activities). The type and number of outputs should reflect a CoP's priorities and the agreed level of engagement by its members, in the work plan, key outputs would be identified for each objective with a designated lead (person responsible for its completion) and timelines.

Monitoring Policy implementation: A CoP may decide to monitor progress in Policy implementation, discuss progress and identify measures to strengthen implementation. The CoP may propose minor updates to Policy; however if a new

Policy is required, POWERGRID would mandate a Policy team to undertake the following actions:

(1) Identifying Policy Coherence For Development (PCD) Priorities. An important role for CoPs is to identify key areas for PCD engagement. Priority areas may emerge because of external factors (response to government priorities; events) or recognition of an area being of key importance for whole of government engagement. Identified priorities should be communicated to the Policy Coherence for Development Coordinator.
(2) IMPROVING PRACTICE. A CoP can promote and support best practice in the planning, implementation, review and evaluation through discussions of programme challenges and successes, or through the preparation of fact sheets and knowledge notes.
(3) Sharing information. Information sharing is a critical role of a CoP and activities to facilitate updating staff (including at posts) on good practice, and relevant or emerging issues should be included in its work plan.

5.2.11.4 Reporting the Progress

A template had been enclosed for reporting the progress made by various CoPs of POWERGRID every month electronically and through hard copy to chief manager (HRD) Knowledge Management Cell, Corporate HRD.

5.2.12 Training on Knowledge Management

POWERGRID had formalized the training on Knowledge Management in its Learner's Planner 2011–12 with the following purpose and content for its Middle Management.

'The Programme aims to provide participants with insight on Knowledge Management with all innovative processes, applications and practices, potential issues, so that measuring Knowledge Management asset is achieved through step by step implementation in POWERGRID's context' (POWERGRID 2011b, p. 69). The contents highlighted that it will be 'Inter-relatedness of information and knowledge, existing technology, purpose of Knowledge Management' (POWERGRID 2011b, p. 69). POWERGRID repeated this training in 2012 for the Knowledge Management Nodal Officers also (POWERGRID 2012, p. 19). POWERGRID repeated the said programme for its Middle Management again in 2013 in order to reinforce the Knowledge Management Implementation in Company (POWERGRID 2013, p. 43). The Company was serious to treat Knowledge Management as an ongoing process and again conducted two programmes for top executives in the year 2014 (POWERGRID 2014a, b, p. 49). Training has been taken very seriously by the management of POWERGRID which has been exhorting and motivating the employees to continuously upgrade their skill, ability and aptitude towards the contribution to Knowledge Bank and use the knowledge available in Knowledge Bank for their performance so that cost could be reduced and time could be saved by avoiding the 'reinvention of wheel'.

5.2.13 *Structure*

Knowledge Management has been visualized as a virtual organization in which every individual staff has been allowed to either contribute or use the knowledge available in Knowledge Bank and its repositories. So far as, the coordination of Knowledge Management activities is concerned, every department, region, RLDC has got a Nodal Officer to coordinate the activities related with the Knowledge Management. Domain Leaders have also been identified who are volunteers and who have been adding value to the contributed knowledge piece. Every Head of the Department has been empowered to take a decision whether the contributed knowledge piece by Affinity Member should be made public or not. This has been decided so that the secret knowledge could be retained for competitive advantage. With the advent of Knowledge Bank and its availability to all personnel of POWERGRID the hierarchical pattern of structure has practically been demolished and it has made the Knowledge Management to come into being in real sense of the term. The Knowledge Management in POWERGRID has been supported by stated organizational objectives and values. Knowledge Management Practices and activities have been complied in line with the values of Company and Corporate Governance.

5.2.14 *Top Management Support*

In a big organization like POWERGRID managers themselves work interdependently bringing their various strengths to the mix. The Top Management has been devoting its energy as the leaders are reminding POWERGRID about its critical role having created the vision and the narrative and looking out for the health of the system. The five key indicators are right motivation, curiosity, insight, engagement and determination have been the hallmark of Top Management of POWERGRID. The Top Management of POWERGRID has been asking various questions and they are innovative leaders because POWERGRID is the only Company in the world which have 1200 kV of transmission line. This is an outcome of the belief of Top Management that all the personnel feel that they are a part of the community having shared purpose, supporting core values, generating ideas through candid discourse and debate. The Top Management has also shown the ability to test ideas through quick pursuit, reflection and adaptation. In the context of Knowledge Management, Top Management support has been demonstrated through the ability to make integrated decisions. POWERGRID has also made it imperative that 'all Heads of

Department shall ensure that this Policy is operationalized within their respective Departments. All employees shall be trained and coached (Mentored by Head of Department and the Knowledge Management Cell to implement this Policy)' (POWERGRID 2010, p. 33).

5.2.15 Reward, Incentive and Recognition

POWERGRID has instituted the following as part of its incentive, reward and recognition for knowledge work.

(1) 'Contribution to KM may be included as a key performance indicator in performance appraisal system of POWERGRID.
(2) Knowledge Currency Unit (KCU) scheme may be established to incentivize sharing of tacit knowledge. A list of employees with KCUs earned will be made on the KM web page. KCU may be used as a virtual currency to award the contributors at the end of the calendar year. One contribution of the knowledge to Knowledge Management Cell may attract 50 KCU. KCU scheme shall include Domain Leaders besides Affinity Members. The virtual currency of 50 KCU shall also be awarded to the user of KM portal who reuse the knowledge from the Knowledge Bank and post a feedback about it. For every feedback, the original contributor will get additional 10 KCU.
(3) List of top contributors with their total KCUs will be published on Knowledge Portal for each month.
(4) The Affinity Members, Domain Leaders and User of KM portal may be conferred titles as Gurus, Reviewer, Knowledge Amateurs etc. to create enthusiasm and peer pressure which is the prime mover behind any new scheme to evangelize it.
(5) A 'Rolling Shield' shall be given to the 'Best community of practice' which is based on Knowledge Management activities done during the calendar year.
(6) Sharing of explicit knowledge as per knowledge map defined by team of nodal officers will be mandatory for the designated person and can be reflected in his/her KRA' (POWERGRID 2010, p. 27).

5.2.16 Mentor/Mentee Programme

POWERGRID implemented Mentor/Mentee Programme to allow the sharing of tacit knowledge from experts to novices. Mentoring is a powerful approach by one person, with relevant experience, helps another person, with lesser experience, to do their job more effectively, and progress in their career. While mentoring is often provided in an informal manner, organizations are recognizing the power of establishing a formal approach to mentoring to help emerging talent gain a solid foothold in the organization. POWERGRID launched this programme with following contour:

- Objectives of the programme:
  - The right approach to mentoring,
  - Building trust credibility and respect with the mentee,
  - Understanding responsibilities of a mentor.
- Primary Competency Category:
  - Human Resource Management—manages the process of aligning human capital with organizational goals.
  - Initiative—proactively makes things happen, evaluates self and others and takes positive corrective action and is self-disciplined.
  - Related competency categories:
  - Professionalism—Projects an image of maturity and integrity that creates credibility.
  - Accountability—Demonstrates personal responsibility and holds self and others accountable for outcomes that are in alignment with the direction of the organization.

In order to implement the programme, POWERGRID organized two workshops, namely (1) Workshop on Mentor Development, and (2) Workshop on Mentee Development during fiscal 2012–2013. The objectives of these programmes were as follows: (1) by the end of the programme, participants will be able to demonstrate themselves as competent mentors by facilitating effective mentoring process as prescribed by the relevant Policy, so as to enhance supportive work culture of performance and commitment; and (2) by the end of the programme, participant will be able to demonstrate themselves as competent mentees by facilitating effective mentoring process as prescribed by the relevant Policy, so as to enhance supportive work culture of performance and commitment.

The contents of the above two workshops are given below:

(1) Workshop on Mentor Development

  (i) Launching an initiative,
  (ii) Building employee engagement,
  (iii) Creating a partnership,
  (iv) Building trust, credibility and respect with mentees,
  (v) Delegation to mentees and
  (vi) Enhancing measurable success, etc.

(2) Workshop on Mentee Development

  (i) Interpersonal competence (Connect with others)
  (ii) Team Building basics,
  (iii) Being a contributing team member,
  (iv) Time management and
  (v) Foundation for success, etc.

The executive category of employees who were experts had been identified as mentors under the mentoring scheme and the executives who required mentoring were identified as mentees under the Mentor–Mentee Scheme. One 'Workshop on Mentor Development' (10–11 December 2012) and two 'Workshop on Mentee Development' (29–30 November 2012 and 20–21 December 2012) were conducted by 'Dale Carnegie Training'.

5.2.17 Library and Information Centre

POWERGRID has established a Library and Information Centre where they have stacked not only books but also have taken access to knowledge resources of various repositories such as ProQuest, Elsevier, ABSCO, Emerald, Sage online, Harvard publication and other search engines through which access could be had to various research and development materials to enhance the knowledge of its personnel. POWERGRID has developed its own library which is primarily intended for use for all its departmental employees. It contains a wealth of information in the field of Operation and Maintenance of transmission assets, Telecom, Grid Management, Knowledge Management, Engineering (Electrical, Mechanical and Civil) besides books on human resource management, finance, information and communication technology and ERP. The library of POWERGRID also subscribes to various journals from all over the globe. In addition to this research reports, proceedings of various workshops and conferences, public documents, Indian Standards, SOI Toposheets, etc. are available in the Library for ready references.

5.2.18 Futuristic Steps

POWERGRID has decided to obtain ISO 25006 certification for making POWERGRID as Knowledge Management compliant Company. It has also been aiming to apply for Most Admired Knowledge Enterprise (MAKE) Award. With these two futuristic steps, POWERGRID aims to have Knowledge Management implemented in all its facets, especially to gain the recognition and increasing its brand equity apropos KM in the power Industry and eco system.

5.2.19 Evaluation of Knowledge Management

POWERGRID has been evaluating the performance on the front of Knowledge Management by Knowledge Audit. It has also been taking into consideration the number of hits on Knowledge Bank. Though Corporate Yellow Pages of POWERGRID besides Knowledge Maps and Matrices have been uploaded and its usage could be gauged by the number of hits to these documents.

5.3 Content Analysis

The content analysis of the texts of Focus Group discussion, interview and elite interview had been imported to NVIVO 10 software for qualitative analysis, and the same has been structured under the heads of word, length, count, weighted percentage and similar words as given in Table 5.1.

In order to get more clarity by observing the precision and easy inference, Word Frequency Query of major words was performed on the data and the same is given in Table 5.2.

Tables 5.1 and 5.2 and the following representative Figs. 5.2 and 5.3 depict that 'knowledge' has got highest percentage of 9.02 which connotes that the environment of KM in POWERGRID is overwhelming. The environment and the ambiance have been created, and KM has been implemented successfully. HRM-related tools such as communication, change management, work activities, individual employee's attitude are predominantly present in the Company. Contrary to it, technology has got the frequency of 0.16 which indicates that technology is subservient to HR factors and thus remains and enabler only. It reinforces that Human Resource Department of POWERGRID has to ensure KM in the Company.

With a view to allow the focus Word Frequency Query, the cloud Fig. 5.2 has been given. The figure clearly brings out that knowledge as a resource is the central theme and the organization (POWERGRID) has accepted it as a management tool. The inter-relatedness of Knowledge Management in POWERGRID has been communicated throughout the organization and these activities have been undertaken as a part of change management. With a view to give a bird's-eye view of the clusters of words, Fig. 5.3 is delineated. The qualitative tools of research have been used for exploratory purposes.

5.4 Quantitative Analysis of Interpretations and Results

The quantitative research will draw on the description of the multiple variables of the study. To complete the study properly, it is necessary to analyse the data collected in order to test the hypotheses and answer the research questions. As already indicated in the preceding chapters, data are interpreted in the descriptive and analytical forms.

This chapter comprises the analysis, presentation and interpretation of findings, results from the study. This phase of the analysis is based on the results of the Questionnaire which has dealt with a quantitative analysis of the data.

Table 5.1 NVivo result

Word	Length	Count	Weighted percentage (%)	Similar words
2010	4	1	0.06	2010
2012	4	2	0.11	2012
2013	4	1	0.06	2013
2014	4	6	0.34	2014
25006	5	1	0.06	25006
Ability	7	18	0.29	Ability, acquisition, breadth, competence, depth, innovation, resource, sense, system, vigilance, vision
Abundance	9	1	0.06	Abundance
Accountability	14	7	0.08	Accountability, assess, finance, reading, review
Accuracy	8	1	0.06	Accuracy
Acquire	7	37	0.29	Acquire, acquiring, acquisition, address, adopt, available, capitalize, capture, derive, developing, development, feel, find, gain, gainfully, get, locate, make, obtain, produce, produced, share, source, take
Acquisition	11	5	0.03	Acquisition, addition, contribution, offering, present
Across	6	3	0.17	Across
Actions	7	40	0.39	Actions, change, contact, decision, designation, determination, dispatch, economy, exchange, execution, executive, flow, following, force, fulfil, going, increase, mechanism, modification, progress, reference, respect, satisfaction, specialization, speed, taking
Activities	10	157	1.54	Actions, activities, activity, alignment, analysis, attempt, best, business, capture, care, categorization, classification, collation, conduct, consultancy, continuance, contribution, cover, development, drawing, enterprise, existing, feedback, flow, formation, formulation, going, heating, help, job, keeping, last, lead, literature, maintenance, measure, measuring, mechanism, open, operation, organization, orientation, part, participants, place, planning, post, practice, presentation, process, project, reading, reducing, research, review, role, routine, search, second, services, share, source, specialization, support, supporting, technology, test, third, training, tuning, undertaking, usage, use, work, workshop
Address	7	13	0.09	Address, addresses, available, cover, covered, place, present, reading, reference
Adequate	8	6	0.15	Adequate, adequately
Advantage	9	8	0.08	Advantage, favourably, gain, lead, positive
Affinity	8	2	0.03	Affinity, related
Afterwards	10	2	0.09	Afterwards, subsequent
Agencies	8	4	0.08	Agencies, agency', means, ways
Alignment	9	6	0.04	Address, alignment, coordination, focus
Also	4	7	0.40	Also, besides

(continued)

Table 5.1 (continued)

Word	Length	Count	Weighted percentage (%)	Similar words
Among	5	3	0.17	Among
Anand	5	1	0.06	Anand
Anchors	7	1	0.06	Anchors
Annual	6	3	0.07	Annual, year, years
Anomalies	9	1	0.06	Anomalies
Another	7	1	0.06	Another
Appended	8	1	0.06	Appended
Appropriate	11	3	0.05	Appropriate, appropriately, proper
Approved	8	8	0.07	Approved, authorized, cover, recognition, reference, support
Area	4	22	0.14	Area, areas, capital, cell, centre, core, countries, keep, open, place, study, well
Articles	8	2	0.06	Articles, lead
Arun	4	2	0.11	Arun
Assets	6	46	0.93	Advantage, asset, assets, bank, capital, contribution, gain, gate, help, keep, lead, maintenance, means, part, resource, security, share, support
Associate	9	26	0.29	Associate, associating, combine, Company, connectivity, director, identify, link, member, name, player, related
Attempt	7	12	0.08	Attempt, best, contribution, part, seek, share, test, undertake, undertaking
Attend	6	5	0.03	Associate, attend, present, second, serve
Authorized	10	21	0.20	Approved, authorized, capital, certification, clearly, day, declare, generate, generation, name, official, post, professional, source, support, sure, white
Available	9	11	0.16	Availability, available, free, help, helped, open, services
Awareness	9	8	0.10	Aware, awareness, feel, orientation, sense
Base	4	6	0.08	Base, based, bases, means, themes, third
Basically	9	2	0.06	Base, basically
Basis	5	5	0.07	Base, based, bases, basis
Bazzwords	9	1	0.06	Bazzwords
Become	6	18	0.09	Become, come, follow, get, going, make, proper, take, work
Benefits	8	3	0.04	Advantage, benefits
Best	4	2	0.02	Best, betterment
Betterment	10	2	0.02	Betterment, major
Beyond	6	1	0.06	Beyond
BHEL	4	1	0.06	BHEL
Blogs	5	1	0.06	Blogs
Board	5	6	0.05	Board, boards, get
Book	4	13	0.18	Annual, book, books, capital, fine, glossary, job, keep, records, reference
Brought	7	1	0.06	Brought

(continued)

Table 5.1 (continued)

Word	Length	Count	Weighted percentage (%)	Similar words
Build	5	6	0.03	Build, centre, formed, make, making
Bureaucratic	12	1	0.06	Bureaucratic
Business	8	44	0.46	Business, businesses, concerned, contribution, field, finance, generation, job, leveraging, lines, literature, official, place, post, services, specialization, technology, undertaking, work
Capabilities	12	9	0.19	Activity, adequate, capabilities, competence, potential
Capital	7	5	0.08	Capital, capitalize, capitalized, means
Captured	8	15	0.26	Appropriate, appropriately, capture, captured, exchange, get, work
Care	4	9	0.10	Attend, aware, care, elaborate, particular, sure
Case	4	9	0.11	Case, example, faced, time, types
Categorization	14	21	0.27	Assessment, categorization, classification, classify, coordination, evaluation, group, number, review
Cell	4	3	0.03	Cell, keep, rod
Centre	6	7	0.08	Capital, centre, centres, core, hearted
Challenge	9	8	0.06	Challenge, hard, process
Change	6	176	1.63	Acquire, adapt, addition, address, adopt, become, betterment, blue, build, capitalize, capture, change, changing, combine, come, completion, content, contract, core, cover, cross, culture, depart, development, elaborate, even, exchange, experience, feel, find, flow, focus, follow, following, formation, fructify, fulfil, gain, get, give, going, help, implement, incorporate, increase, initiation, initiative, innovation, installation, integrate, journey, know, level, load, make, map, meet, mobile, modification, modify, number, obtain, open, organization, part, people, post, prime, process, produce, progress, provide, refine, represent, satisfaction, specialization, speed, spread, step, still, subject, synthesize, take, test, time, top, total, trace, translate, visit, white, work, write, yellow
Chauhan	7	2	0.11	Chauhan
Circulars	9	2	0.05	Circulars, global
Circulation	11	9	0.14	Circulars, circulate, circulation, disseminate, dissemination, mobile, spread
Clearly	7	8	0.06	Broad, clearly, focus, free, open
Collaboration	13	1	0.06	Collaboration
Collaboratory	13	1	0.06	Collaboratory
Collate	7	2	0.03	Collate, collation
Collected	9	30	0.28	Book, case, collation, collected, collections, collective, content, core, field, findings, information, mail, number, post
Combine	7	8	0.12	Combine, incorporate, integrate, one, synthesize
Come	4	6	0.05	Address, come, derive, future, present

(continued)

Table 5.1 (continued)

Word	Length	Count	Weighted percentage (%)	Similar words
Comment	7	5	0.07	Comment, inputs, noticed, pipeline, reference
Committee	9	6	0.11	Board, committee
Common	6	3	0.06	Common, informal, shared
Communication	13	157	1.73	Addition, address, analysis, audio, book, brief, challenge, circulation, comment, common, communication, communities, contact, content, core, cover, description, discussion, dissemination, document, drawing, fact, formulation, get, give, giving, identification, indicate, indicator, information, interview, keep, lead, level, link, literature, mail, material, matter, measure, moderator, national, network, number, offering, opinion, papers, place, post, present, presentation, programme, project, reading, recognition, reference, report, represent, request, result, sense, share, sharing, source, step, subject, term, test, time, topic, trace, transmission, usage, video, view, visit, write
Company	7	19	0.18	Companies, Company, mover, number
Competence	10	11	0.13	Adequate, competence, field, place
Competitive	11	6	0.10	Challenge, competitive, competitor, meet, open
Completion	10	12	0.09	Completion, dispatch, fulfil, implement, top, total
Compliant	9	1	0.06	Compliant
Conceptualized	14	5	0.05	Conceptualized, find
Concerned	9	26	0.25	Business, care, concerned, least, load, matter, part, reference, related
Conducive	9	4	0.08	Conducive, contributed, contribution, lead
Conducted	9	6	0.11	Conduct, conducted, conducting, manner, respect
Connectivity	12	20	0.27	Articulate, communication, connectivity, contact, continuance, continuously, cover, embed, know, lead, link, meeting, post, related, serve
Consecutively	13	3	0.10	Consecutively, successful, successfully
Consider	8	2	0.03	Consider, counts
Consultancy	11	4	0.04	Consultancy, reference
Contains	8	19	0.15	Bank, case, classify, contains, gate, incorporate, measure, moderator, still, well
Contents	8	143	1.20	Address, agenda, area, book, business, centre, certification, challenge, comment, competitive, content, contents, contract, core, culture, description, domain, example, experience, fact, feedback, field, front, indicator, information, latest, lead, learning, list, major, material, matter, offering, opinion, outline, part, particular, pilot, plan, programme, project, purpose, reading, recognition, reference, region, report, request, result, satisfaction, sense, source, specification, study, subject, target, technology, term, topic, trace, value, values, view, whole
Contextual	10	1	0.06	Contextual

(continued)

Table 5.1 (continued)

Word	Length	Count	Weighted percentage (%)	Similar words
Continuance	11	19	0.13	Continuance, continuously, cover, covered, cross, follow, keep, make, spread, take, trace, usage, use, value, visit
Contract	8	18	0.25	Contract, contracts, declarations, declare, Policy, provide, reducing, undertake, undertaking
Contributed	11	6	0.05	Combine, contributed, contribution, give, giving, offering
Cops	4	2	0.11	Cops
Core	4	3	0.03	Core, effectiveness, effects
Corporate	9	12	0.26	Collected, collections, collective, corporate, incorporate, material
Cost'	5	12	0.19	Assessment, cost', cover, fine, maintenance, satisfaction, terms
Countries	9	7	0.33	Bangladesh, bhutan, countries, ethiopia, india, national
Courses	7	11	0.14	Courses, flow, formed, lines, orientation, steps, workshop
Cover	5	30	0.27	Bank, blue, case, change, cover, covered, cross, flow, grit, hop, overall, prime, spread, take, test, top, white
Creating	8	95	1.16	Adopt, attempt, build, capture, coil, conduct, cover, create, created, creating, cross, derive, elaborate, elicit, establish, facilitate, find, follow, force, found, free, fructify, fulfil, generate, get, give, incorporate, initiate, keep, lead, make, making, map, organize, outline, pioneer, place, plan, present, produce, produced, programme, project, reference, renew, represent, result, serve, support, time, trace, undertake, work, write
Crucial	7	4	0.09	Crucial, material, pivotal
Customer	8	5	0.06	Customer, regular, usage, use
Day	3	5	0.08	Day, even, year, years
Decision	8	9	0.07	Crucial, decision, designation, determination, measure, recognition, step
Declarations	12	20	0.29	Contract, declarations, declare, explicit, expressed, find, indicate, indicator, take, undertake
Dedicated	9	5	0.08	Dedicated, give, giving, security
Defines	7	6	0.06	Defines, determination, name, outline, setting, specified
Delineated	10	12	0.21	Capitalize, cross, defines, delineated, drawing, lines, outline, represent, specified, trace, write
Department	10	19	0.37	Depart, department, departments, going, part, personnel, security, started
Departmental	12	2	0.11	Departmental
Designated	10	42	0.38	Address, advantage, cross, designated, designation, exchange, identification, indicate, indicator, innovation, innovations, mrs, name, number, patterns, place, planning, points, post, purpose, recognition, second, specified, term, view

(continued)

Table 5.1 (continued)

Word	Length	Count	Weighted percentage (%)	Similar words
Desirous	8	13	0.24	Care, desirous, like, please, seek, wanted
Determination	13	34	0.18	Adopt, assess, determination, field, follow, identify, know, measure, name, place, purpose, regulations, setting, support, take, time, value
Developing	10	29	0.24	Build, come, culture, derive, developing, development, elaborate, find, follow, formulated, make, progress
Different	9	10	0.16	Change, depart, different, gap, spread, various
Directly	8	55	0.41	Address, addresses, chairman, conduct, conducted, conducting, directives, directly, focus, give, lead, level, make, orientation, pilot, place, points, project, send, subject, take, taking, target, training, ways, work
Discussions	11	15	0.32	Address, consider, cover, discussion, discussions, initiate, interview, lead, recognition
Dispatch	8	3	0.04	Dispatch, hits, remove
Document	8	26	0.19	Book, brief, certification, contract, document, documents, measure, opinion, papers, process, report, source, specification, study, well
Domains	7	42	0.59	Area, areas, domain, domains, field, front, major, region, responsibility, strategy, study, subject, technology
Done	4	3	0.17	Done
Drawing	7	21	0.10	Drawing, force, get, make, making, outline, plan, project, represent, study, trace
Drivers	7	1	0.06	Drivers
Dwivedi	7	4	0.23	Dwivedi
Effectiveness	13	11	0.07	Dispatch, effectiveness, effects, facilitate, force, fulfil, get, help, serve
Efficiency	10	5	0.15	Economy, effectiveness, effects, efficiency
Elaborate	9	6	0.05	Elaborate, outline, refine, represent
Electronically	14	1	0.06	Electronically
Emerged	7	3	0.06	Emerged, future
Employees	9	18	0.84	Employee, employees, mover
Enabling	8	6	0.29	Cover, enabled, Enablers, enables, enabling
Enriching	9	1	0.06	Enriching
Enrolled	8	1	0.06	Enrolled
Ensure	6	9	0.16	Cover, ensure, make, secured, security
Enterprise	10	16	0.11	Business, collective, enterprise, initially, initiate, initiation, initiative, operation
Entrusted	9	2	0.06	Entrusted, give
Environment	11	23	0.31	Area, bank, case, context, domain, environment, field, front, responsibility, setting
Envisaged	9	9	0.10	Envisaged, project
Equipments	10	12	0.14	Base, equipments, material, monitor, prepared, still, target, white
Especially	10	5	0.14	Especially, exception, particular, specialization

(continued)

Table 5.1 (continued)

Word	Length	Count	Weighted percentage (%)	Similar words
Establishing	12	38	0.33	Bank, base, based, bases, build, Company, effectiveness, effects, establish, established, establishing, found, give, giving, headquarters, installation, make, making, mover, name, official
Etc	3	2	0.11	Etc.
Evaluation	10	62	0.56	Adopt, assess, assessment, bank, believe, capitalize, challenge, classify, consider, declare, evaluation, feel, find, identify, include, know, like, make, measure, measures, measuring, place, project, respect, review, test, value, values, view
Even	4	4	0.03	Even, level, regular, still
Every	5	3	0.17	Every
Example	7	7	0.13	Example, exception, lessons, models, pilot, represent
Execution	9	14	0.19	Ceo, conduct, execution, executive, give, lead, make, mechanism, practice, step
Existing	8	16	0.18	Come, day, existing, fact, flow, obtain, potential, present, real
Expects	7	2	0.06	Believe, expects
Experience	10	21	0.15	Come, experience, experiences, feel, find, know, meet, test, time, vision
Experiential	12	1	0.06	Experiential
Experts	7	19	0.73	Expert, expertise, experts, practice, practices, professional, technical
Exposure	8	2	0.03	Exposure, still
Expressed	9	25	0.30	Articulate, comment, declare, explicit, expressed, faced, field, give, indicate, measure, present, represent, step, time
External	8	8	0.16	External, internal, international, outside
Faced	5	13	0.08	Base, bottom, faced, front, lines, present, presentation, top, undertake
Facilitate	10	9	0.29	Facilitate, facilitated, facilitates, facilitating, help, helped
Factors	7	5	0.22	Agents, component, components, factors
Failure	7	3	0.08	Failure, force, outage
FAQS	4	1	0.06	FAQS
Favourably	10	6	0.07	Advantage, favourably, following
Feel	4	14	0.11	Care, experience, experiences, feel, opinion, respect, satisfaction, security, Sense
Files	5	2	0.07	Document, files
Finance	7	8	0.07	Bank, finance, leveraging
Find	4	46	0.31	Acquire, address, base, certification, designation, determination, feel, find, findings, get, identification, locate, map, noticed, number, obtain, place, sense, support, take, total, trace, translate
Flow	4	8	0.13	Circulate, courses, flow, ripple
Focus	5	6	0.06	Centre, centres, focus

(continued)

Table 5.1 (continued)

Word	Length	Count	Weighted percentage (%)	Similar words
Following	9	17	0.25	Adopt, adopted, adopting, come, cover, follow, followed, following, next, regular, succeeded, trace
Force	5	15	0.12	Affinity, Company, force, headquarters
Formal	6	6	0.07	Formal, official, positive
Formation	9	11	0.06	Bank, bottom, formation, initially, initiate, initiation, initiative
Formed	6	22	0.22	Base, core, description, formed, gap, grid, like, make, making, manner, measure, patterns
Formulated	10	10	0.18	Articulate, conceptualized, developing, development, expressed, formulated, formulating, formulation
Forums	6	1	0.06	Forums
Found	5	8	0.08	Found, initially, initiate, initiation, initiative, innovation, innovations
Free	4	6	0.07	Available, free, give, sparing
Friendly	8	3	0.04	Associate, friendly, informal
Front	5	6	0.08	Base, faced, far, front
Fructify	8	2	0.02	Fructify, setting
Functional	10	11	0.10	Formal, functional, functioning, functions, hop, initiation, installation, official, serve
Future	6	4	0.07	Future, next, succeeded
Gain	4	23	0.17	Benefits, capitalize, clearly, derive, find, gain, gainfully, get, hits, make, making, take, top
Garg	4	1	0.06	Garg
Gave	4	1	0.06	Gave
Generate	8	8	0.07	Free, generate, generation, give, giving, make, returns
Get	3	28	0.18	Come, drawing, experience, experiences, force, generate, generation, get, give, implement, lead, make, making, process, serve, started, take
Give	4	53	0.40	Appropriate, articulate, bank, board, bottom, capitalize, coal, combine, contribution, finance, free, fulfil, gate, get, give, giving, help, make, making, meet, offering, open, place, present, presentation, provide, represent, serve, share, step, support, thank, top
Global	6	3	0.10	Global, globalization, international
Goals	5	14	0.17	Business, goals, purpose, target, view
Going	5	55	0.37	Address, change, circulate, come, cover, cross, despatch, dispatch, follow, going, hop, journey, last, lead, make, offered, offering, open, post, process, progress, ripple, sail, seek, serve, speed, spread, started, step, take, top, trace, well, work
Got	3	4	0.23	Got

(continued)

Table 5.1 (continued)

Word	Length	Count	Weighted percentage (%)	Similar words
Group	5	112	1.09	Alignment, bank, blue, book, business, categorization, classification, collation, Company, content, core, culture, domain, economy, embed, enterprise, field, findings, following, force, formation, free, front, generation, group, groups, information, initiate, leaders, mail, meeting, network, organization, people, personnel, post, structure, system, team, year
Groupware	9	1	0.06	Groupware
Guidelines	10	1	0.06	Guidelines
Harnessing	10	1	0.06	Harnessing
Hearted	7	2	0.03	Hearted, respect
High	4	3	0.03	Broad, high, last
Hits	4	9	0.05	Bottom, get, hits, make, top
Hours	5	2	0.04	Hours, prime
However	7	3	0.12	However, still
Human	5	15	0.37	Business, human, literature, metrics
Identify	8	16	0.23	Identified, identify, identifying, known, list, name, number
Immense	7	1	0.06	Immense
Implementation	14	28	0.92	Core, execution, executive, give, implement, implementation, implemented, implementing, means, practice, rod, step
Improving	9	21	0.28	Addition, become, betterment, build, care, development, fructify, help, improving, maintenance, progress, refine
Incentivize	11	1	0.06	Incentivize
Includes	8	7	0.20	Cover, include, includes, including, incorporate, initiate
Incorporate	11	6	0.09	Incorporate, increasing, integrate, integrated
Increasing	10	21	0.29	Addition, build, combine, gain, gainfully, give, include, increase, increasing, maximizing, modify, progress, speed, spread, synthesize
Indeed	6	1	0.06	Indeed
Indicate	8	14	0.13	Case, identification, indicate, indicator, level, points, reading, reference, step, suggested, target, trace
Individual	10	100	1.47	Associate, best, broad, case, chairman, chief, Company, competitor, customer, employee, executive, expert, failure, help, identification, individual, initiate, intellectual, job, major, manager, material, member, moderator, monitor, mover, name, national, official, owner, persons, pioneer, planner, player, possible, professional, provider, regular, researcher, respective, self, sir, source, subject, target, user, various, white

(continued)

Table 5.1 (continued)

Word	Length	Count	Weighted percentage (%)	Similar words
Information	11	63	0.61	Agenda, book, brief, bulletin, case, certification, circulate, comment, common, contents, cover, despatch, dispatch, disseminate, elaborate, example, exception, fact, friendly, give, indicate, indicator, informal, information, latest, lead, list, material, particular, pipeline, post, present, presentation, programme, reading, report, represent, respect, send, source, specific, spread, translate
Initially	9	17	0.12	Address, attempt, cover, establish, found, initially, initiate, initiation, initiative, installation, name, open, organize, pioneer, started, undertake
Innovation	10	3	0.04	Innovation, innovations, pioneer
Inputs	6	10	0.08	Business, inputs, load
Installation	12	20	0.18	Bank, base, field, grid, headquarters, heating, installation, network, post, source
Internal	8	6	0.11	Global, incorporate, internal, international, national
Interview	9	7	0.09	Consultancy, interview
Inviting	8	9	0.15	Challenge, inviting, process
ISO	3	3	0.17	ISO
Issues	6	25	0.41	Change, effectiveness, effects, emerged, free, issues, outcome, published, result, results, returns, take, taking
January	7	1	0.06	January
Job	3	4	0.04	Job, problems
Journey	7	3	0.04	Journey, sail, visit
Keeping	7	19	0.17	Book, classify, continuance, continuously, cover, document, follow, gate, keep, keeping, maintain, post, still, take
Kept	4	1	0.06	Kept
Know	4	16	0.11	Awareness, feel, find, know, orientation, recognition, sense, take
Knowers'	8	1	0.06	Knowers'
Knowledge	9	275	9.02	Ability, acquisition, area, breadth, case, categorization, centre, classification, cognitive, content, core, culture, domain, example, exception, experience, fact, field, grid, information, initially, initiate, initiation, initiative, innovation, know, knowledge, lead, learning, level, matter, open, operation, opinion, organization, orientation, particular, place, plan, practice, process, programme, purpose, reading, recognition, region, respect, search, sense, specific, structure, study, subject, system, target, topic, values, view, vision
Kumar	5	2	0.11	Kumar
Kumud	5	3	0.17	Kumud
Kunaujia	8	1	0.06	Kunaujia
Lead	4	5	0.03	Chairman, give, lead, major, moderator

(continued)

Table 5.1 (continued)

Word	Length	Count	Weighted percentage (%)	Similar words
Leaders	7	10	0.17	Chairman, chief, director, executive, leaders, manager, moderator, monitor
Learning	8	36	0.28	Acquire, acquiring, acquisition, determination, find, learning, practice, reading, study, take, taking, test, work
Learnt	6	1	0.06	Learnt
Level	5	14	0.11	Depth, even, force, high, level, low, points, setting, stages, top
Libraries	9	1	0.06	Libraries
Like	4	12	0.19	Affinity, care, like, please, potential
Lines	5	29	0.32	Contact, front, lines, link, name, part, people, pipeline, Policy, steps, support, trace, tuning
List	4	9	0.09	Agenda, contents, list, name, post
Load	4	3	0.05	Load, money
Locate	6	62	0.39	Acquire, address, area, base, bottom, capital, cell, centre, core, depth, development, domain, environment, field, focus, front, gap, high, lead, locate, locations, map, meeting, open, outside, part, place, post, region, setting, source, steps, target, top, work
Low	3	3	0.04	Blue, low
Mail	4	3	0.03	Mail, post, send
Make	4	25	0.15	Attempt, dispatch, establish, facilitate, force, found, fulfil, get, help, initiate, make, making, name, organize, pioneer, serve, time, undertake
Management	10	111	2.45	Build, care, certification, chairman, conduct, conducting, directives, directly, director, directors, finance, give, lead, make, management, manager, managing, meet, name, organization, organize, possible, process, work
Managerial	10	1	0.06	Managerial
Manner	6	3	0.03	Manner, respect, ways
Manuals	7	1	0.06	Manuals
Map	3	10	0.22	Functional, functioning, functions, map, maps, represent
Material	8	12	0.16	Base, blue, crucial, formulation, indicator, material, network, real, sail, temporal
Matrices	8	3	0.17	Matrices
Matter	6	25	0.26	Area, counts, issues, least, matter, produce, subject, topic
May	3	2	0.11	May
Means	5	21	0.11	Core, get, identify, list, means, name, number, plan, purpose, reference, represent, sense
Measure	7	35	0.33	Care, information, last, length, measure, measures, measuring, metrics, reading, security, step, steps, test, time
Mechanism	9	12	0.14	High, low, mechanism, mobile, second, third, transmission

(continued)

Table 5.1 (continued)

Word	Length	Count	Weighted percentage (%)	Similar words
Meeting	7	32	0.32	Alignment, bank, case, contact, cover, cross, embed, flow, front, fulfil, group, meet, meeting, meetings, serve, team, Visit
Member	6	4	0.09	Director, member, members
Metrics	7	3	0.07	Metrics, money
Minaxi	6	1	0.06	Minaxi
Minutes	7	11	0.17	Hours, minutes, second, time
Mis	3	2	0.11	Mis
Miscellaneous	13	1	0.06	Miscellaneous
Models	6	2	0.04	Models, patterns
Modify	6	5	0.05	Load, modify, positive
Name	4	9	0.18	Designation, mrs, name, reference, specification, term, viz
Nodal	5	7	0.40	Nodal
Non	3	1	0.06	Non
Noticed	7	3	0.04	Broad, noticed, post
NTPC	4	1	0.06	NTPC
Number	6	43	0.93	Base, circulation, come, counts, five, four, frequency, issues, lead, list, make, number, one, prime, real, routine, routines, ten, thousand, three, total, transmission, two
Objects	7	75	0.89	Addition, area, audio, bank, base, bottom, business, cell, centre, challenge, component, cover, depth, domain, exception, field, focus, formation, gap, installation, length, making, material, mechanism, objectives, objects, outside, part, place, real, region, source, structure, target, top, total, whole
Officers	8	30	0.62	Agencies, agency', authorized, functional, functioning, functions, headquarters, major, officers, part, place, placed, positive, post, role
Official	8	7	0.09	Authorized, established, formal, official, regular
Open	4	10	0.08	Available, establish, found, gap, initiative, open, possible, spread
Operation	9	23	0.18	Addition, cover, functional, functioning, functions, gate, operation, operations, pilot, procedures, search, support, times, work
Organization	12	220	3.91	Alignment, associate, bank, best, blue, business, case, cell, chief, collective, combine, committee, Company, competitor, coordination, core, corporate, cross, directives, directly, employee, enterprise, establish, established, establishing, executive, expert, failure, five, force, formation, formed, friendly, headquarters, help, individual, initiate, integrated, intellectual, job, management, manpower, material, member, mover, national, official, organization, organize, owner, personnel, plan, planner, player, possible, prepared, preparing, professional, regular, researcher, rod, self, source, structure, structured, subject, system, systems, team, user, white

(continued)

Table 5.1 (continued)

Word	Length	Count	Weighted percentage (%)	Similar words
Organizational	14	5	0.29	Organizational
Orientation	11	4	0.03	Orientation, points, view
ORM	3	1	0.06	ORM
Outside	7	6	0.09	External, open, outside
Owners	6	3	0.13	Owner, owners
Pages	5	6	0.29	Pages, spread
Part	4	52	0.46	Addition, area, audio, base, basis, bottom, capital, coal, component, components, depth, length, making, material, member, open, outline, outside, part, particular, process, region, regional, sail, structure, support, system, top
Participants	12	3	0.07	Participants, player
Particular	10	5	0.04	Particular, specialization, specific, specification, specified
PER	3	2	0.11	PER
Perceived	9	13	0.09	Attend, feel, find, get, perceived, sense, trace
Persons	7	11	0.11	Identification, individual, part, persons, role
Pilot	5	3	0.07	Pilot, piloted
Place	5	59	0.52	Address, capital, centre, collate, directives, directly, even, front, high, identified, identify, identifying, job, know, lead, level, load, local, locate, locations, organize, place, placed, points, positive, post, send, setting, step, structure, take, target, top
Planning	8	53	0.61	Agenda, book, designated, designation, organize, outline, plan, planning, Policy, prepared, preparing, programme, programs, project, strategy, system, time
PNBC	4	1	0.06	PNBC
Points	6	51	0.32	Address, advantage, centre, core, Dubai, exchange, focus, front, high, job, last, lead, meeting, mobile, place, points, post, respect, second, setting, source, target, term, time, top, work, workshop
Policy	6	10	0.13	Base, Policy
Positive	8	38	0.29	Address, alignment, cover, declare, formal, formation, front, last, level, list, locate, locations, orientation, place, positive, presentation, respect, second, step, sure, take, target, terms, third, view
Possible	8	11	0.32	Base, basis, day, given, possible, potential
POWERGRID	9	36	2.07	POWERGRID
Powergridians	13	1	0.06	Powergridians
Practice	8	26	0.36	Consultancy, execution, executive, follow, functional, much, operation, operations, patterns, possible, practice, practices, review, usage, use, used
Prashant	8	1	0.06	Prashant
Pratyksha	9	1	0.06	Pratyksha
Precious	8	4	0.10	Precious, value, values, wanted

(continued)

Table 5.1 (continued)

Word	Length	Count	Weighted percentage (%)	Similar words
Prepared	8	35	0.33	Developing, development, formulated, formulating, formulation, keep, make, making, planning, practice, prepared, preparing, prime, provide, review, setting, spread, training, work
Present	7	17	0.15	Existing, exposure, indicate, place, present, presentation, programme, represent, stages, support
Prints	6	3	0.04	Prints, step, trace
PRM	3	1	0.06	PRM
Problems	8	8	0.08	Case, matter, problems
Procedures	10	8	0.12	Functional, functioning, functions, procedures, routine, routines
Process	7	73	0.85	Acquisition, actions, assessment, awareness, categorization, classification, coordination, culture, determination, developing, development, identification, know, learning, operation, operations, planning, prepared, procedures, process, processes, progress, reading, recognition, refine, routine, search, serve, study, work
Produce	7	7	0.07	Director, keep, lead, produce, produced, result
Programs	8	8	0.09	Job, pilot, programme, programs
Progress	8	14	0.13	Build, hop, increase, increasing, progress, top
Project	7	39	0.44	Business, designated, designation, enterprise, envisioning, external, map, operation, plan, planning, project, projects, undertake, undertaking
Promoting	9	5	0.08	Advantage, help, lead, promoting
Proper	6	4	0.04	Appropriate, proper, real, specific
Provide	7	27	0.37	Articulate, board, bottom, capitalize, coal, fulfil, gate, give, help, meet, offered, offering, open, provide, provided, provider, provides, serve, step, top
Published	9	6	0.13	Contribution, prints, published, write
POWERGRID	9	1	0.06	POWERGRID
Queries	7	4	0.09	Queries, request, seek, test
Rare	4	2	0.11	Rare
Ravi	4	12	0.69	Ravi
Reading	7	3	0.03	Reading, records, trace
Real	4	4	0.04	Fact, proper, real
Records	7	22	0.27	Audio, book, cover, document, fine, keep, list, maintain, minutes, post, records, review, take, video, write
Reducing	8	8	0.07	Contract, low, reducing, top
Redundant	9	2	0.09	Redundant, sparing
Reference	9	24	0.15	Annual, give, glossary, identify, list, name, number, reference, source, subject, target
Refreshing	10	5	0.26	New, refreshing
Regional	8	29	0.40	Area, areas, centre, common, department, development, domain, environment, field, open, place, region, regional, setting

(continued)

Table 5.1 (continued)

Word	Length	Count	Weighted percentage (%)	Similar words
Regular	7	15	0.11	Even, official, organization, regular, regulations, system
Regulations	11	2	0.03	Coordination, regulations
Related	7	10	0.05	Alignment, formation, list, name, place, related
Relevant	8	1	0.06	Relevant
Remove	6	3	0.08	Far, remove
Renew	5	5	0.06	Capitalize, change, exchange, renew
Reports	7	22	0.62	Accountability, bulletin, comment, cover, covered, despatch, dispatch, papers, pipeline, report, reports, study, themes
Represent	9	22	0.12	Capture, contact, make, map, outline, present, project, represent, support
Request	7	19	0.13	Book, challenge, interview, keep, place, process, request, seek
Required	8	23	0.91	Expects, must, need, needed, needs, required, requirement, requirements, requires, requisite, take, taking, wanted
Researcher	10	28	0.50	Research, researcher, search, test
Resources	9	17	0.40	Help, keep, maintenance, resource, resources, support
Respect	7	6	0.08	Report, respect, respective, respectively, value, values
Responsibility	14	16	0.27	Accountability, feedback, job, keeping, requirement, respect, responsibility
Results	7	13	0.25	Attend, come, decision, follow, lead, outcome, result, results, solutions, subsequent
Retiring	8	1	0.06	Retiring
Returns	7	4	0.03	Get, returns, trace
Reuse	5	3	0.14	Reuse, reused, reusing
Review	6	5	0.10	Comment, refreshing, review, reviews
Revised	7	2	0.11	Revised
Robust	6	1	0.06	Robust
Role	4	17	0.15	Functional, functioning, functions, part, purpose, role, second, third, use, used
SAARC	5	1	0.06	SAARC
Scared	6	1	0.06	Scared
Search	6	4	0.03	Feel, search, seek
Security	8	7	0.08	Certification, secured, security, share, sure
Send	4	8	0.09	Despatch, dispatch, mail, send, transmission
Sense	5	9	0.10	Perceived, sense, view, vision
Serve	5	10	0.05	Bank, facilitate, help, represent, serve
Services	8	5	0.04	Functional, serve, services, setting
Sessions	8	1	0.06	Sessions
Setting	7	31	0.25	Company, core, domain, field, focus, group, hard, setting, time
Shankar	7	1	0.06	Shankar

(continued)

Table 5.1 (continued)

Word	Length	Count	Weighted percentage (%)	Similar words
Sharing	7	27	0.56	Common, contributed, contribution, giving, offering, part, share, shared, sharing
Shri	4	22	1.26	Shri
Since	5	2	0.11	Since
Singh	5	15	0.86	Singh
Sometimes	9	1	0.06	Sometimes
Southern	8	1	0.06	Southern
Sparable	8	1	0.06	Sparable
Specialists	11	2	0.07	Specialists, specialization
Specific	8	5	0.05	Particular, proper, specific, specification, specified
Specified	9	4	0.04	Provide, specific, specified
Spread	6	8	0.08	Circulation, dissemination, prepared, spread
Stages	6	4	0.05	Boards, generation, setting, stages
Started	7	18	0.12	Adopt, formation, initiation, initiative, innovation, installation, open, organization, started
Strategically	13	1	0.06	Strategically
Structure	9	49	0.37	Area, base, boards, case, cell, centre, coil, cross, functional, grid, integrate, integrated, keep, level, organization, pilot, plan, sail, still, structure, structured, study, support, system, ways
Study	5	21	0.15	Collate, consider, follow, major, map, name, pioneer, research, review, search, study, trace, view
Subject	7	40	0.42	Capabilities, case, content, contents, field, give, literature, major, national, open, strategy, study, subject, technology, themes, topic
Succeeded	9	14	0.13	Following, make, next, succeeded, successful, successfully, test, work
Successful	10	5	0.21	Made, successful, successfully
Sunita	6	2	0.11	Sunita
Support	7	39	0.43	Associate, attend, base, based, care, cover, document, documents, ensure, establish, facilitate, friendly, help, helped, keep, keeping, lead, maintenance, post, provide, second, serve, source, step, support, supporting
System	6	15	0.21	Codify, economy, link, network, programme, structure, system, systems
Tacit	5	9	0.52	Implicit, tacit
Take	4	40	0.37	Acquire, acquiring, adopt, adopted, adopting, appropriate, base, board, capture, conduct, consider, contains, contract, contracts, field, follow, free, get, make, making, name, reading, remove, take, taking, test, use
Taken	5	1	0.06	Taken
Teams	5	7	0.28	Five, team, teams
Techniques	10	1	0.06	Techniques
Technology	10	9	0.16	Technical, technologies, technology
Telecom	7	5	0.25	Telecom, video

(continued)

Table 5.1 (continued)

Word	Length	Count	Weighted percentage (%)	Similar words
Terms	5	14	0.10	Subject, term, terms, time
Therefore	9	3	0.17	Therefore, thus
Though	6	1	0.06	Though
Throughout	10	2	0.11	Throughout
Time	4	13	0.10	Coordination, day, lead, prime, regular, second, time, times
Took	4	2	0.11	Took
Tools	5	2	0.11	Tools
Top	3	11	0.06	Capital, clearly, cover, covered, hop, lead, top
Topic	5	5	0.07	Local, topic
Transformer	11	20	0.24	Become, betterment, coil, come, process, transformer, translate, work
Transmission	12	11	0.13	Circulation, contract, contracts, dissemination, giving, transmission
Tyagi	5	1	0.06	Tyagi
Updation	8	4	0.23	Updated, update
Upload	6	1	0.06	Upload
Use	3	45	0.36	Address, board, consultancy, development, functional, give, help, implement, practice, purpose, reuse, role, share, take, technology, usage, use, used, work
User	4	2	0.08	User, users
Vadhwa	6	1	0.06	Vadhwa
Various	7	9	0.18	Different, individual, respect, respective, respectively, various
Vendor	6	2	0.11	Vendor
Verticals	9	6	0.29	Hierarchical, post, verticals
View	4	20	0.18	Capitalize, consider, decision, determination, exposure, identify, include, like, make, opinion, respect, value, view
Vision	6	5	0.10	Vision, visions
Voluntarily	11	1	0.06	Voluntarily
Wadhwa	6	2	0.11	Wadhwa
Ways	4	6	0.14	Gate, interstate, portal, steps, ways
Websites	8	3	0.11	Portal, Websites
Well	4	5	0.04	Advantage, well
Went	4	1	0.06	Went
Whole	5	71	0.71	Area, board, build, case, cell, coil, completion, core, cover, cross, entire, field, future, gap, grid, implement, individual, innovation, installation, intact, integrate, integrated, lead, level, load, material, means, mechanism, network, one, sail, source, step, structure, system, test, top, total, trace, ways, well, whole
Without	7	1	0.06	Without

(continued)

Table 5.1 (continued)

Word	Length	Count	Weighted percentage (%)	Similar words
Work	4	81	0.47	Analysis, bank, book, capital, care, coil, enterprise, exchange, formed, functional, functioning, functions, going, help, job, make, making, number, operation, operations, part, place, project, reference, represent, research, review, role, search, serve, services, source, spread, still, study, take, undertaking, use, work, workshop
Workarounds	11	1	0.06	Workarounds
Workshop	8	2	0.07	Workshop, workshops
Write	5	46	0.30	Address, adopt, analysis, book, brief, capitalize, certification, cross, development, document, execution, job, lead, literature, matter, measure, opinion, outline, papers, place, reference, report, review, source, specification, study, well, write
Yatindra	8	4	0.23	Yatindra
Yes	3	2	0.11	Yes

On the basis of availability of data, descriptive analytical and factor analyses have been done. The sample size of the study is 414 taken from the various levels of executives of POWERGRID, who are rationally aware about the Knowledge Management in all its facets across the Company.

On the basis of Questionnaire, multiple variables have been created and factors have been reduced.

5.4.1 Demographic Description

The levels of executives from E1 to E9 may be referred to where E1 is the junior most executive and E9 is senior most.

Table 5.3 and Fig. 5.4 reflect the cross-tabulation of gender and age of the respondents. Total number of male executives are 370 and females are 44. Maximum number of male in POWERGRID exists between the ages of 47–54, i.e. 97 males. Maximum number of females exists between the ages of 23–30, i.e. 13 females. Aggregate number of executives exists between the ages of 47–54, i.e. total 106 persons.

Table 5.4 and Fig. 5.5 reflect the cross-tabulation of age and education of the executives in POWERGRID. Out of total 240 graduate executives, maximum number of graduates falls in the age between 23 and 30, i.e. 35, and out of 168 postgraduates, maximum number of postgraduates falls in the age between 47 and 54, i.e. 45. Similarly out of total 6 executives having their qualification as doctorate,

Table 5.2 Word frequency query

Knowledge	Organization	POWERGRID	Activities	Creating	Implement	Process	Officers	Sharing	Business	Issues	Locate	Designation		Take	Dept.	Provide	Going
	Management	Communication	Individual	Group	Required	Employees	Reports	Evaluation	Part	Give	Human	Prepare	Measure	Countries	Points	Meeting	Lines
		Change	Shri	Number	Objects	Experts	Planning	Place	Project	Also	Structure	Possible	Express	Pages	Acquire	Increase	Facilitate
			Contents	Assets	Singh	Whole	Information	Tacit	Support	Nodal	Practice	Discuss	Vertical	Associated	Declaration	Collected	Learning
						Ravi	Domains	Research	Subject	Resource	Use	environment	enabling	Teams	Cover	Categories	Response
								Work	Directly	Regional	2014	Find	Ability	Improving	Matter	Capture	Results
										Actions	Establish	Write	Org.	Records	Corporate	Concentrate	Telecom
													Positive	Connection	Refresh	Following	Setting

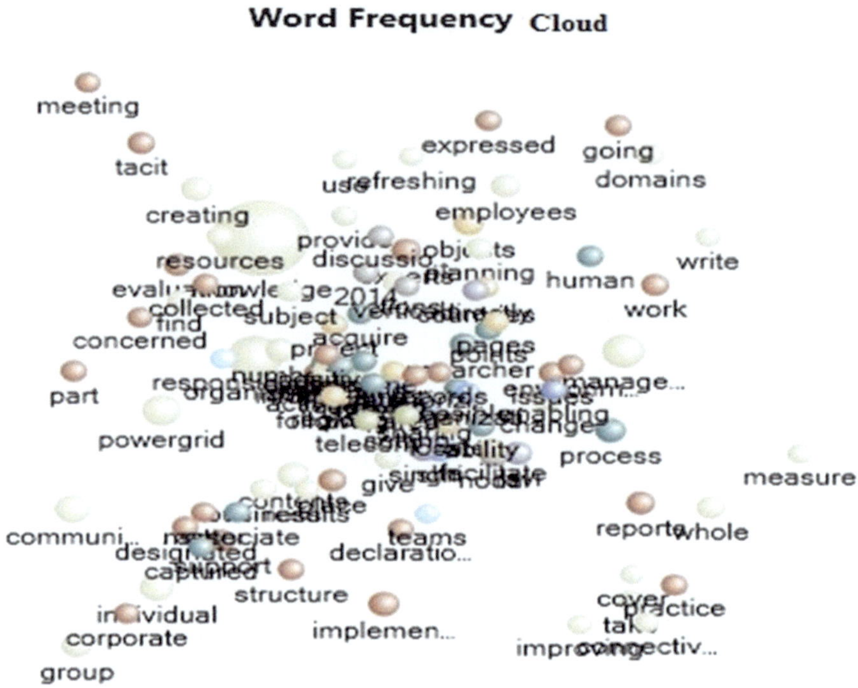

Fig. 5.2 Word frequency cloud

maximum doctorate executes falls in the age group of 47–54, i.e. 3. From the point of view, maximum number of qualified executives comes under the age group of 47–54, i.e. 106.

Table 5.5 reflects the cross-tabulation to define department-wise executive levels (E1–E3, E4–E5, E6–E7, E8 and above) and their educational qualification. Maximum number of educationally qualified executives are in the departments which are clubbed as 'Enablers' with defined executive levels, i.e. 145, while consultancy having least number of educationally qualified executives, i.e. 36. In the Operation and Maintenance Department, out of total 83 executives, maximum number of executives is graduates, i.e. 16, and only 15 executives are postgraduates with defined executive level. In the Grid Management Department, out of total 44 executives, maximum number of executives is graduates, i.e. 29, 13 executives are postgraduates while 2 executives are having doctorate qualification with defined executive level. In the Project Management Department, out of total 68 executives, maximum number of executives is graduates, i.e. 51, and only 17 executives are postgraduates with defined executive level.

Fig. 5.3 Frequency layout

Table 5.3 Gender and age

Gen * age cross-tabulation						
Count						
Gen	Age					Total
	23–30	31–38	39–46	47–54	55 and above	
Male	81	80	75	97	37	370
Female	13	11	11	9	0	44
Total	94	91	86	106	37	414

Similarly, in the Telecom Department, out of total 38 executives, maximum number of executives is graduates, i.e. 25 and only 13 executives are postgraduates with defined executive level. In the Consultancy Department out of total 36

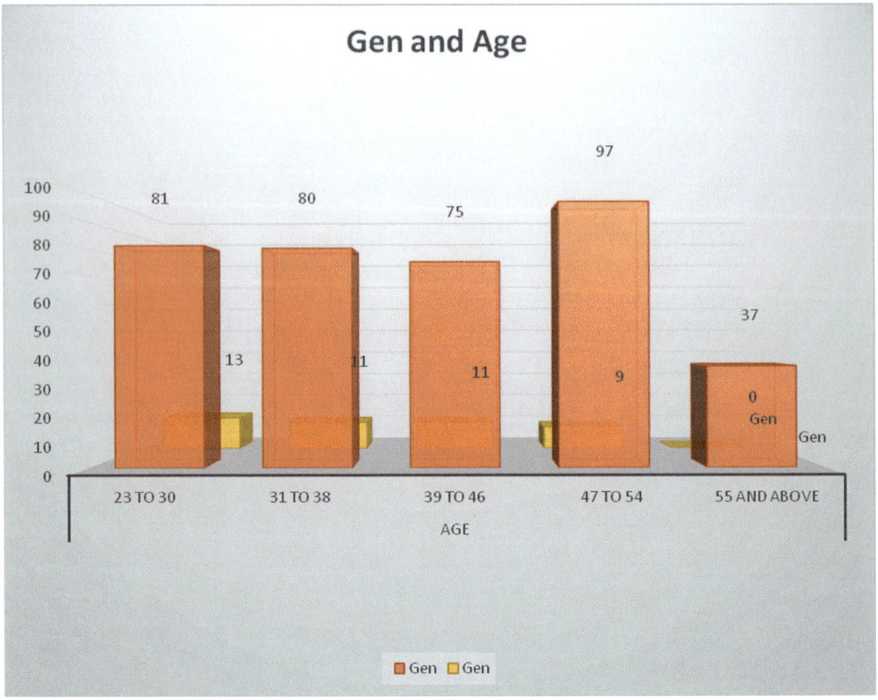

Fig. 5.4 Age and education

Table 5.4 Age and education

Age * Edu cross-tabulation				
Count				
Age	Edu			Total
	Graduation	Postgraduation	Doctorate	
23–30	65	29	0	94
31–38	58	32	1	91
39–46	42	42	2	86
47–54	58	45	3	106
55 and above	17	20	0	37
Total	240	168	6	414

executives, maximum number of executives is graduates, i.e. 16, and only 20 executives are postgraduates with defined executive level. Finally in Enablers category, out of total 145 executives, maximum number of executives is graduates,

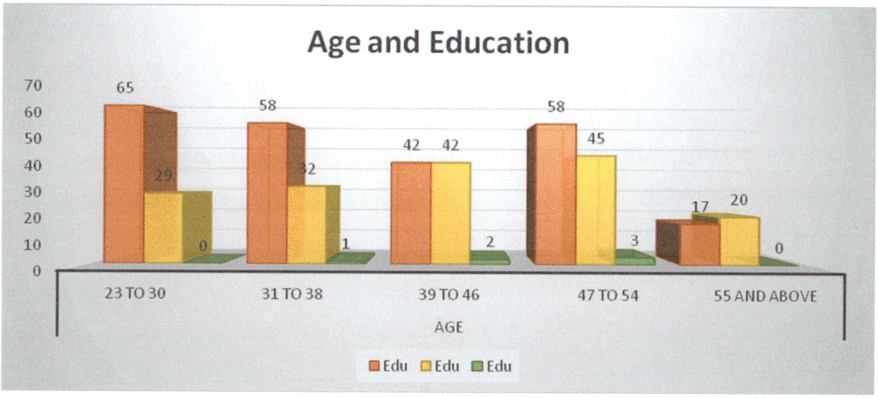

Fig. 5.5 Age and education

i.e. 51, 90 executives are postgraduates while 4 executives have having doctorate qualification with defined executive level.

Finally, out of aggregate figure of 414 executives, maximum number of executives is graduates, i.e. 240, 168 executives are postgraduates while only 6 executives are having doctorate qualification with defined executive level. Therefore, we can say that the maximum number of executives in POWERGRID is graduates. Some executives are postgraduates while only few executives in POWERGRID have doctorate qualification (Table 5.5).

Table 5.6 reflects the cross-tabulation of executives associated with Knowledge Management Work, Education and defined level of executives in POWERGRID. In the defined levels, total 172 executives fall in E1–E3 out of these only 61 accepts to be associated with Knowledge Management Work and 111 have denied their association with Knowledge Management Work. Out of 123 from the levels of E4–E5, 43 accepted their association with Knowledge Management Work and 80 denied. From the level of E6–E7 level out of 110, 56 accepted their association with Knowledge Management Work and 54 denied. Similarly in total, 9 executives fall in the level of E8 and above, and out of these 5 accepted their association with Knowledge Management Work and 4 denied.

In aggregate, out of all 414 respondents 165 executives are associated with Knowledge Management in which 56 are graduate, 75 are postgraduates while 4 executives are having doctorate qualification. From the 249 respondents who denied their association with Knowledge Management Work, 154 are graduates, 93 are postgraduates while 2 executives are having doctorate qualification.

The period of joining POWERGRID by executives of both genders who participated in survey has been elaborated countwise in Table 5.7 (Fig. 5.6, Tables 5.8 and 5.9).

Table 5.5 Level, education and department

Level * Edu * Dept cross-tabulation					
Count					
Dept		Edu			Total
		Graduation	Postgraduation	Doctorate	
Operation and maintenance	E1–E3	36	4		40
	E4–E5	21	4		25
	E6–E7A	9	7		16
	E8 and above	2	0		2
	Total	68	15		83
Grid management	E1–E3	18	5	1	24
	E4–E5	6	3	0	9
	E6–E7A	5	4	1	10
	E8 and above	0	1	0	1
	Total	29	13	2	44
Project management	E1–E3	21	5		26
	E4–E5	12	4		16
	E6–E7A	17	7		24
	E8 and above	1	1		2
	Total	51	17		68
Telecom	E1–E3	6	3		9
	E4–E5	13	4		17
	E6–E7A	6	6		12
	Total	25	13		38
Consultancy	E1–E3	7	7		14
	E4–E5	4	3		7
	E6–E7A	5	10		15
	Total	16	20		36
Enablers	E1–E3	29	30	0	59
	E4–E5	13	35	1	49
	E6–E7A	8	22	3	33
	E8 and above	1	3	0	4
	Total	51	90	4	145
Total	E1–E3	117	54	1	172
	E4–E5	69	53	1	123
	E6–E7A	50	56	4	110
	E8 and above	4	5	0	9
	Total	240	168	6	414

Table 5.6 KMW and education

KMW * Edu * level cross-tabulation						
Count						
Level			Edu			Total
			Graduation	Postgraduation	Doctorate	
E1–E3	KMW	Yes	35	26	0	61
		No	82	28	1	111
	Total		117	54	1	172
E4–E5	KMW	Yes	23	20	0	43
		No	46	33	1	80
	Total		69	53	1	123
E6–E7A	KMW	Yes	26	26	4	56
		No	24	30	0	54
	Total		50	56	4	110
E8 and above	KMW	Yes	2	3		5
		No	2	2		4
	Total		4	5		9
Total	KMW	Yes	86	75	4	165
		No	154	93	2	249
	Total		240	168	6	414

5.5 Measurement, Reliability and Validity of Major Constructs

To test the relation among the 81 variables on the basis of theoretical model of hypotheses and further constructs have been framed. Again those factors have been divided in the group and nonparametric tests have been done for testing the hypothesis. In many social science disciplines, it is common practice to 'convert' ordinal Likert or Likert-style scale data into interval data by assigning numbers, such as '1' for 'Strongly Disagree' or '5' for 'Strongly Agree'. Among researchers, statisticians and evaluators, this practice is controversial. However, it is widely used.

A number of articles argued or assumed that Likert items do not form an interval scale, but instead should be considered ordinal scales and should be analysed accordingly (e.g. Coombs 1960; Vigderhous 1977; Jakobsson 2004; Jamieson 2004; Knapp 1990; Kuzon et al. 1996). Other articles proposed ways to get around this perceived ordinal/interval scale 'problem' by proposing alternative Likert-like item formats such as the two-stage alternative offered by Albaum (1997) or the phrase completion alternative offered by Hodge and Gillespie (2003).

To get the appropriate result of hypothesis, nonparametric tests have been done.

Through confirmatory factor analysis, all the factors have been confirmed whether they are coming under the category of predefined model or not. This is being elaborated in the coming sections.

Table 5.7 Level year of joining and gender

Current level in POWERGRID * year of joining * gender cross-tabulation

Count

Gender		Year of joining																												Total
		1986	1987	1988	1989	1990	1991	1992	1993	1994	1995	1996	1997	1998	1999	2000	2001	2002	2003	2004	2005	2006	2007	2008	2009	2010	2011	2012	2013	
Male	E1–E3	0	1	0	1	2	12	3	3	2	3	2	0	0	0	1	0	0	1	4	3	13	13	7	24	20	10	3	21	149
	E4–E5	1	1	0	1	1	24	5	5	4	3	0	1	5	2	12	4	11	4	10	8	1	0	1	3	2	0	0	2	111
	E6–E7A	0	1	1	2	1	44	9	18	7	4	9	1	2	0	0	0	0	0	0	0	0	0	0	0	0	0	0	1	100
	E8 and above	0	0	0	0	0	6	2	1	0	0	0	0	0	0	0	0	0	0	0	0	0	0	0	0	0	0	0	0	9
	Total	1	3	1	4	4	86	19	27	13	10	11	2	7	2	13	4	11	5	14	11	14	13	8	27	22	10	3	24	369
Female	E1–E3				2	0	1	1	0	0			0			1	0		0	1	1	2	1	1	6	3	1		2	23
	E4–E5				0	1	1	1	1	0			1			1	1		1	0	3	0	0	0	1	0	0		0	12
	E6–E7A				0	0	4	0	2	2			0			1	0		0	0	1	0	0	0	0	0	0		0	10
	Total				2	1	6	2	3	2			1			3	1		1	1	5	2	1	1	7	3	1		2	45
Total	E1–E3	0	1	0	3	2	13	4	3	2	3	2	0	0	0	2	0	0	1	5	4	15	14	8	30	23	11	3	23	172
	E4–E5	1	1	0	1	2	25	6	6	4	3	0	2	5	2	13	5	11	5	10	11	1	0	1	4	2	0	0	2	123
	E6–E7A	0	1	1	2	1	48	9	20	9	4	9	1	2	0	1	0	0	0	0	1	0	0	0	0	0	0	0	1	110
	E8 and above	0	0	0	0	0	6	2	1	0	0	0	0	0	0	0	0	0	0	0	0	0	0	0	0	0	0	0	0	9
	Total	1	3	1	6	5	92	21	30	15	10	11	3	7	2	16	5	11	6	15	16	16	14	9	34	25	11	3	26	414

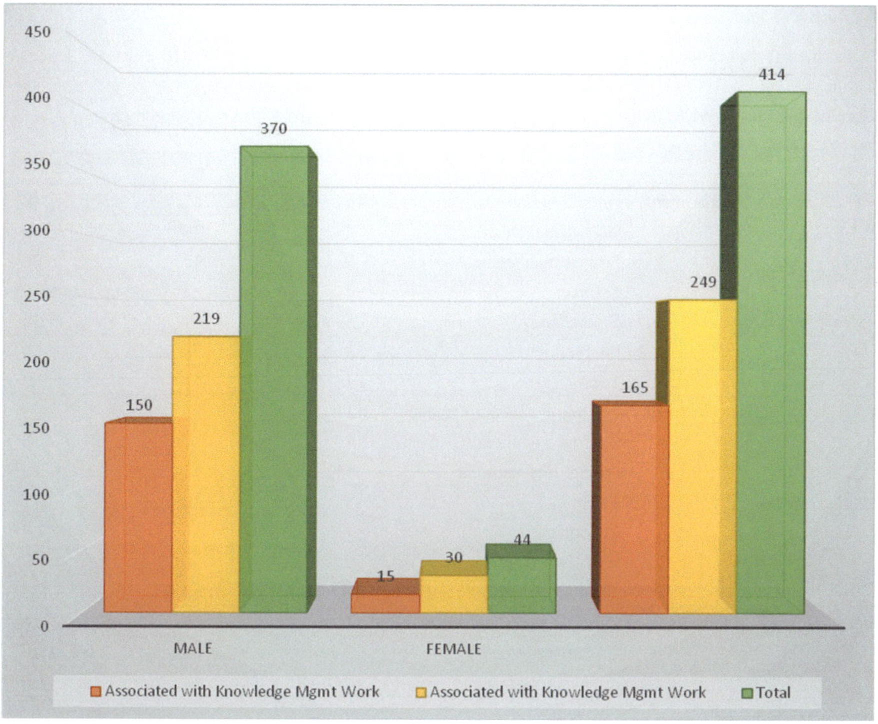

Fig. 5.6 Gender-wise association with KM

5.5.1 *Factor Analysis*

Table 5.10 reflects the results of KMO for sample adequacy and Bartlett's test of sphericity for this section.

5.5.2 *Reason of Doing Exploratory Factor Analysis (EFA)*

All the variables are differentiated in the specific group of Knowledge Management Implementation. Groups are formed on the basis of the related variables then hypothesis tests have been done. Even though it was required to go for EFA because questions were randomly selected and designed. There are no such model and theory have been developed in field of KM. Until or unless statistically analysed constructs cannot be defined it was very difficult to give the direction to the research. These are the strong reasons for exploring of constructs from the variables.

Table 5.8 Dept-wise association with KM

Department * associated with knowledge mgmt. Work * current level in POWERGRID cross-tabulation					
Count					
Current level in POWERGRID			Associated with knowledge mgmt. work		Total
			Yes	No	
E1–E3	Department	Operation and maintenance	16	24	40
		Grid management	4	20	24
		Project management	9	17	26
		Telecom	4	5	9
		Consultancy	5	9	14
		Enablers	23	36	59
		Total	61	111	172
E4–E5	Department	Operation and maintenance	11	14	25
		Grid management	2	7	9
		Project management	5	11	16
		Telecom	5	12	17
		Consultancy	1	6	7
		Enablers	19	30	49
		Total	43	80	123
E6–E7A	Department	Operation and maintenance	11	5	16
		Grid management	3	7	10
		Project management	10	14	24
		Telecom	4	8	12
		Consultancy	9	6	15
		Enablers	19	14	33
		Total	56	54	110
E8 and above	Department	Operation and maintenance	1	1	2
		Grid management	1	0	1
		Project management	1	1	2
		Enablers	2	2	4
		Total	5	4	9
Total	Department	Operation and maintenance	39	44	83
		Grid management	10	34	44
		Project management	25	43	68
		Telecom	13	25	38
		Consultancy	15	21	36
		Enablers	63	82	145
		Total	165	249	414

KMO (Kaiser-Mayer-Olkin): Kaiser-Mayer-Olkin (1974) recommends accepting values greater than 0.5 as acceptable (values below this should lead you to either collect more data or rethink which variables to include). Furthermore, values between 0.5 and 0.7 are mediocre, values between 0.7 and 0.8 are good, values between 0.8 and 0.9 are great and values above 0.9 are superb (Hutcheson and Sofroniou 1999, pp. 224–225). For these data, the value is 0.93, which falls into

Table 5.9 Gender-wise association with KM

Gender * associated with knowledge mgmt. work cross-tabulation				
Gender		Associated with knowledge mgmt. work		Total
		Yes	No	
Male	Count	150	220	370
	% of total	36.2 %	53.1 %	89.4 %
Female	Count	15	29	44
	% of total	3.6 %	7.0 %	10.6 %
Total	Count	165	249	414
	% of total	39.9 %	60.1 %	100.0 %

Table 5.10 KMO and Bartlett's

KMO and Bartlett's test		
Kaiser–Meyer–Olkin measure of sampling adequacy		0.953
Bartlett's test of sphericity	Approx. chi-square	18794.132
	df	2926
	Sig.	0.000

the range of being superb: so, we should be confident that factor analysis is appropriate for these data. KMO measured of sampling adequacy mixed the minimum criteria, the value of KMO measure of sampling adequacy for this set of variable is 0.953 which could be labelled as marvellous.

Bartlett's Test of Sphericity: Bartlett's test is another indication of the strength of the relationship among variables. This is also the way to test the null hypothesis. For factor analysis to work, we need some relationships between variables and if the R-matrix were an identity matrix, then all correlation coefficients would be zero. Therefore, we want this test to be significant (i.e. have a significance value less than 0.05). A significant test tells us that the R-matrix is not an identity matrix; therefore, there are some relationships between the variables. For these data, Bartlett's test is highly significant ($p < 0.001$), and therefore, factor analysis is appropriate.

5.5.3 Pattern Matrix

The pattern matrix has been calculated on oblique rotation. Oblique rotation is that the factors are allowed to correlate. This rotation was done because there were highly related variables in the data. It is strongly believed that all the factors of KM in the organization can be implemented effectively when they are correlated. KM implementation can be done only by collectively using the constructs. Table 5.11 reflects the pattern matrix of multiple variables. Rotation converged in 21 iterations

Table 5.11 Result of factor analysis on oblique rotation

Pattern matrix[a]

	Factor														
	1. IT-enabled KM activities	2. Recognition and contribution of best facilities	3. Intranet and Internet platforms	4. Ergonomic facilities for KM	5. Communications are facilitated	6. Top Management support	7. Colloquium and seminar	8. KM in actions	9. Unused knowledge assets	10. KM: a way of life	11. Uses of knowledge tools	12. Periodical recognition for contribution	13. Learning from experience	14. Collation of knowledge and its capitalization	15. KM-oriented mindset
Cronbach's alpha	0.89	0.888	0.839	0.711	0.753	0.856	0.799	0.745	0.767	0.745	0.82	0.856	0.772	0.754	0.765
% of variances	34.325	3.678	3.550	3.037	2.522	2.162	1.863	1.760	1.691	1.605	1.567	1.543	1.465	1.414	1.340
Eigen values	26.430	2.832	2.733	2.339	1.942	1.665	1.435	1.355	1.302	1.236	1.207	1.188	1.128	1.089	1.032
V-66	0.704														
V-58	0.662														
V-59	0.637														
V-45	0.633														
V-46	0.553														
V-65	0.549														
V-44	0.474														
V-60	0.468														
V-79	0.427														
V-70	0.415														
V-18		0.936													
V-17		0.691													
V-19		0.648													
V-16		0.569													
V-28		0.507													
V-14		0.463													
V-15		0.419													
V-30			0.800												
V-29			0.751												
V-32			0.749												
V-31			0.515												
V-33			0.436												

(continued)

Table 5.11 (continued)

Pattern matrix[a]

	Factor														
	1. IT-enabled KM activities	2. Recognition and contribution of best facilities	3. Intranet and Internet platforms	4. Ergonomic facilities for KM	5. Communications are facilitated	6. Top Management support	7. Colloquium and seminar	8. KM in actions	9. Unused knowledge assets	10. KM: a way of life	11. Uses of knowledge tools	12. Periodical recognition for contribution	13. Learning from experience	14. Collation of knowledge and its capitalization	15. KM-oriented mindset
V-23				0.709											
V-22				0.640											
V-47				0.476											
V-34				0.429											
V-55					0.720										
V-56					0.616										
V-72						0.849									
V-71						0.634									
V-73						0.549									
V-67							0.953								
V-68							0.884								
V-1								0.961							
V-2								0.885							
V-5									0.826						
V-4									0.713						
V-3									0.676						
V-6									0.673						
V-62										0.895					
V-61										0.704					
V-11										0.505					
V-37											0.621				
V-36											0.600				
V-38											0.563				
V-75												0.869			
V-74												0.510			
V-81													0.649		
V-80													0.598		
V-63														0.564	

(continued)

Table 5.11 (continued)

Pattern matrix[a]

	Factor														
	1. IT-enabled KM activities	2. Recognition and contribution of best facilities	3. Intranet and Internet platforms	4. Ergonomic facilities for KM	5. Communications are facilitated	6. Top Management support	7. Colloquium and seminar	8. KM in actions	9. Unused knowledge assets	10. KM: a way of life	11. Uses of knowledge tools	12. Periodical recognition for contribution	13. Learning from experience	14. Collation of knowledge and its capitalization	15. KM-oriented mindset
V-64														0.517	
V-7															0.542
V-8															0.405

Extraction method: maximum likelihood
Rotation method: promax with Kaiser normalization
[a]Rotation converged in 21 iterations

with the maximum likelihood extraction method implemented. Promax with kappa value 4 have been taken. Absolute value of the coefficient pattern matrix was suppressed below. 15 factors seem to have emerged in the oblique rotation of pattern matrix.

Factor 1: IT-enabled KM activities,
Factor 2: Recognition and contribution of best facilities,
Factor 3: Intranet and Internet platforms,
Factor 4: Ergonomic facilities for KM,
Factor 5: Communications are facilitated,
Factor 6: Top Management Support,
Factor 7: Colloquium and seminar,
Factor 8: KM in actions,
Factor 9: Unused Knowledge assets,
Factor 10: KM: A way of life,
Factor 11: Uses of Knowledge tools,
Factor 12: Periodical recognition for contribution,
Factor 13: Learning from experience,
Factor 14: Collation of knowledge and its capitalization and
Factor 15: KM-oriented mindset.

Some variables were deleted which had low value to get well-structured pattern matrix.

Maximum likelihood component analysis was conducted on the 81 items with oblique rotation (Promax). The KMO measured verified the sampling adequacy for the analysis KMO = 0.953 'superb' (according to Andy 2009) which is well above the acceptable limit of 0.5. Bartlett's test of sphericity $\chi 2$ (2926 = 18974.132, $p < 0.001$) indicated that correlation between item are sufficient large for the analysis. And initial analysis was run to obtain Eigenvalues for each component in the data. 15 components have Eigenvalues over Kaiser's creation of 1 and in combined explained 63.522 of the variance given the large sample size and the covariance of the screen plot and Kaiser's criterion on 15 components. This is the number of components that were retained in the final analysis. Table 5.11 shows the factor loading after rotation.

5.5.4 *Reliability Test*

A total of 15 factors have been extracted in the pattern matrix through maximum likelihood component analysis. Each factor contained certain number of related variables, given specific names. Variables of each factor were tested, reliability test of all the 15 factors have been done. Test of reliability confirms the internal consistency of the variables. Values of Cronbach's Alpha of all the 15 factors are greater than 0.7 which is above the acceptable limit 0.7 (Andy 2009).

% of variance—The '% of variance' row tells that how much of the total variability (in all of the variables together) can be accounted for by each of these summary scales or factors. Factor 1 accounts for 34.325 % of the variability in all 81 variables, and so on.

Eigenvalues—An eigenvalue is equal to the sum of the squared factor loadings for a particular variable on the factor with which the eigenvalue is associated. In simple terms, the larger the eigenvalue, the larger the proportion of variance in the data accounted for by that factor. The plot of eigenvalues against the number of factors (scree plot) was proposed by Cattell (1966) as an aid in deciding on the optimum number of factors to extract. Deciding how many factors will best represent the patterns of correlations in the data is one of the main problems in EFA because although SPSS has a default method for doing this, it is really up to the analyst to decide. By default, the programme will only extract eigenvalues greater than 1.

5.6 Confirmatory Factor Analysis (CFA) Through Structural Equation Modelling (SEM)

In this chapter, the analysis of the data collected based on the frame of reference of this thesis is presented. First, the empirical analysis of the proposed theoretical model using SEM is presented followed by demographic profile of the respondents. The chapter concludes by analysing the demographic, parameter and factors of Knowledge Management. To empirically validate the extended technology accepted model, structural equation modelling (SEM) was used.

The following section briefly describes the factor analysis and introduction to structural equation modelling including the basic concepts of structural equation modelling and moves on to present the psychometric checks done using the measurement model of SEM and the analysis results. Hypotheses tests have been done according to the nature of the available data.

5.7 Concepts of SEM

Structural equation modelling (SEM) is a multivariate technique, which estimates a series of inter-related dependence relationships simultaneously. The term structural equation modelling conveys that the causal processes under study are represented by a series of structural (i.e. regression) equations and that these can be modelled pictorially to enable a clearer conceptualization of the study. The hypothesized model can be tested statistically in a simultaneous analysis of the entire system of variables to determine the extent to which it is consistent with the data. If the goodness of fit is adequate, the model argues for the plausibility of postulated relations among the variables. Given below are some of the basic concepts of SEM and a few terms which are used in the analysis.

5.7.1 Latent and Observed Variables

With regard to the measurement instrument, the variables are classified as latent and observed variables. Latent variables are not observed directly. They are operationally defined in terms of behaviour believed to represent it. The measured scores (measurements) are termed as observed or manifest variables, and they serve as indicators of the underlying construct which they presume to represent. Hence, one latent variable has two or more statements (observed variables) to represent it. Latent variables are represented with eclipses and observed variables are with rectangle.

5.7.2 Factor Analysis and the Development of Model

Factor analysis is one of the oldest and best-known statistical procedures for investigating relationship between sets of observed and latent variables. In using factor analysis, the researcher examines the covariation among a set of observed variables in order to gather information on their underlying latent constructs (i.e. factors).

There are two basic types of factor analysis: exploratory factor analysis (EFA) and confirmatory factor analysis (CFA). The factor analytic model (EFA or CFA) focuses solely on how, and the extent to which, the observed variables are linked to their underlying latent factors. Specifically speaking, it is concerned with the extent to which observed variables are generated by the underlying latent constructs and thus strength of the regression paths from the factors to the observed variables (the factor loadings) is of primary interest. Exploratory factor analysis is designed for situations where links between the observed and latent variables are unknown or uncertain.

Hence, once the items of Questionnaire were finalized, an EFA has been conducted to determine the extent to which the item measurements are related to the latent constructs. In contrast, confirmatory factor analysis (CFA) is used when relations between the observed measures and the underlying factors have been postulated, 'a priori', based on knowledge of the theory, empirical research, or both, and then tests this hypothesized structure statistically. Because the CFA model focuses solely on the link between factors and their measured variables, within the framework of SEM, it represents what is called as a measurement model. In this study, the Questionnaire was developed on the basis of literature review and expert opinion which covers various aspects of KM and the model was developed; hence, both EFA and CFA were used. Either a variance–covariance or correlation data matrix derived from a set of observed or measured variables can be used.

Maximum likelihood estimation is robust to no normality. Parameter estimates are efficient and structural equation (SE) is lightly biased. Chi-square goodness of fit can be quite problematic. But nobody uses chi-square goodness of fit by itself (refer to CFI, TLI, RMSEA). If Bootstrap has been done then there is no normality requirement at all (SE is asymptotic bootstraps) unless you select the 'parametric bootstrap' (João 2014).

The ML (maximum likelihood) estimator is considered relatively robust to violations of normality assumptions (Diamantopoulos et al. 2000). The aim of maximum likelihood estimation is to find the parameter value(s) that makes the observed data most likely. This is because the likelihood of the parameters given the data is defined to be equal to the probability of the data given the parameters. (Technically, they are proportional to each other, but this does not affect the principle.)

Multivariate normality (mvm) is sufficient but not necessary for normal-theory maximum likelihood SEM. There are two counterexamples: Two groups each normally distributed but their means differ. Here, the data from the full sample are non-normal (a mixture of two multivariate normal). Ordinal data are analysed by full-information maximum likelihood. Here, the likelihood is computed by integrating the multivariate normal distribution, so it is normal-theory maximum likelihood but with (e.g.) a threshold model that separates the ordinal categories. Yet the data themselves are decidedly non-normal (Michael 2014).

5.8 Validation of Model and Model Measurement

A confirmatory factor analysis (CFA) was conducted using AMOS 20. Measurement model validity depends on establishing acceptable levels of goodness of fit for the measurement model and finding specific evidence of construct validity. Validity is defined as the extent to which data collection methods accurately measure what they were intended to measure. To satisfy the validity procedure, the following are the validity and reliability checks that were carried out:

- Content validity,
- Convergent validity,
- Composite reliability,
- Discriminant validity and
- Nomo logical validity.

The content validity of the research model has already been presented in previous chapters. The other psychometric property checks of the instrument are presented here.

5.8.1 Convergent Validity

Convergent validity is a method to test construct validity. The word of construct shows a theoretical viewpoint to explain some phenomenon (Wiersma 2000). (Van

Dalen 1973) states that construct usually refers to a complex concept which includes several inter-related factors. In this study, convergent validity was assessed by factor loading, composite reliability (CR) and average variance extracted (AVE) (Fornell and Larcker 1981). Confirmatory factor analysis (CFA) is conducted to estimate factor loading of variables. In fact, a factor loading presents the level of a regression path from a latent to its indicators. According to (Hair et al. 2010), an acceptable factor loading value is more than 0.5 and when it is equal to 0.7 and above it is considered good for one indicator. The level of CR is another guideline to review convergent validity although Cronbach's alpha is a very popular coefficient to test reliability (Bollen and Long 1993; Garson 2011). According to (Hair et al. 2010), the acceptable value of CR is 0.7 and above.

SEM (structural equation modelling) has been done on AMOS 20. Convergent validity was verified for each factor loading. Some data transformations have been done and those observed variables were deleted whose factor loadings were less than 0.7 in the process of model fit analysis (Table 5.12).

In structural equation modelling, for the convergent validity the factor loadings and average variance extracted (AVE) should be greater than 0.5 (Fornell and Larcker 1981). The average variance extracted (AVE) for each of the factors is calculated manually for all the constructs using the formula suggested by (Hair et al. 2010) as given below:

$$\mathrm{AVE} = \frac{\left(\sum_{i=1}^{n} \lambda_i^2\right)}{\left(\sum_{i=1}^{n} \lambda_i^2\right) + \left(\sum_{i=1}^{n} \delta_i\right)}$$

where λ is the standardized factor loadings and δ is the indicator measurement error.

The average variance extracted and the construct factor loadings are presented in Table 5.13. As seen from the table, all AVE values and factor loadings are greater than 0.5 with almost all values above 0.60. For all the constructs, all items have high loadings, with majority above 0.75 therefore demonstrating convergent validity. This study satisfied this criterion; hence, convergent validity was established.

5.8.2 *Composite Reliability*

A requirement for construct validity is score reliability. Reliability can be defined as the degree to which measurements are free from error and, therefore, yield consistent results. Reliability, also called consistency and reproducibility, is defined in general as the extent to which a measure, procedure or instrument yields the same result on repeated trials (Carmines and Zeller 1979). It can be used to assess the degree of consistence among multiple measurements of variables (Hair et al. 2010).

Operationally reliability is defined as the internal consistency of a scale, which assesses the degree to which the items are homogeneous. For reflective measures,

Table 5.12 Construct statements

No.	Construct statement	Standardized factor loadings (λ)
IT-enabled KM activities		
V-66	In my organization, it is mandatory for every employee to document insights (*in the form of best practices, lessons learnt,* etc.) from the work/project	0.76
V-65	In my organization, every employee is encouraged to periodically produce *documents* in the form of technical reports, journal papers, presentations, articles, etc.	0.84
Recognition and contribution of best facilities		
V-17	Innovation and knowledge creation are a way of life in my Company	0.8
V-16	Transfer of knowledge and best practices are taking place in day-to-day working of employees	0.74
Intranet and Internet platform		
V-30	The intranet site provides an online discussion forum in which employees can share information and post their doubts	0.8
V-29	Employees are provided 'instant messenger' or 'chat'—videoconferencing facilities through which they can have conversations with their colleagues	0.75
V-32	Employees are provided the 'blogging' platform wherein they can share their experiences through blogs	0.79
Communications are facilitated		
V-55	All important communications in my organization are sent to me through circulars or memos in electronic environment	0.73
V-56	Noticeboards kept at various vantage points in my organization play an important role in informing me of various news and events	0.89
Top management support		
V-72	My superiors fully endorse my way of handling Knowledge Management-related issues	0.78
V-71	I feel that I have got support from management for Knowledge Management	0.86
V-73	The Top Management supports and encourages Knowledge Management	0.81
Colloquium and seminar		
V-67	My organization sponsors trips for employees to international/national conferences and seminars	0.81
V-68	My organization arranges seminars/lectures by external experts on a periodic basis	0.83
KM in actions		
V-1	Knowledge Management has taken place in POWERGRID	0.74
V-2	In my department, Knowledge Management activities are taking place satisfactorily	0.8

(continued)

Table 5.12 (continued)

No.	Construct statement	Standardized factor loadings (λ)
Uses of knowledge tools		
V-37	I use knowledge maps and matrices of my Company for knowledge requirements	0.85
V-36	I contact subject-matter experts by using my corporate Yellow Pages	0.7
V-38	Our Company has got collegial, flat, team-based, open-door Policy	0.79
Periodical recognition for contribution		
V-74	My organization periodically gives special awards to employees who make significant contributions to the organization	0.84
V-75	Recognition in the form of titles, certificates and commendation letters are periodically awarded to the contributors of knowledge	0.89
Learning from experience		
V-63	Knowledge Management Cell of POWERGRID collates knowledge nuggets and disseminates it through power portal	0.81
V-64	The capitalization of knowledge of Company is realized through our consultancy department	0.75
Collation of knowledge and its capitalization		
V-80	POWERGRID publishes regular intellectual capital reports to its external stakeholders	0.77
V-81	The evaluation of work/projects that are completed successfully has been done in my organization from KM point of view	0.82

Table 5.13 AVE and factor loadings of the factors

Construct	AVE	Construct factor loading
IT-enabled KM activities	0.607	0.8
Recognition and contribution of best facilities	0.648	0.77
Intranet and Internet platform	0.598	0.78
Communications are facilitated	0.597	0.81
Top Management support	0.608	0.82
Colloquium and seminar	0.663	0.82
KM in actions	0.671	0.77
Uses of knowledge tools	0.594	0.78
Periodical recognition for contribution	0.614	0.87
Learning from experience	0.750	0.78
Collation of knowledge and its capitalization	0.630	0.8

all items are viewed as parallel measures capturing the same construct of interest. Thus, the standard approach for evaluation, where all path loadings from construct to measures are expected to be strong (i.e. ≥ 0.70) is used. Composite reliability measures the overall reliability of a set of items loaded on a latent construct. Value ranges between zero and one. Values greater than 0.70 reflect good reliability. Between 0.60 and 0.70 is also acceptable if other indicators of the construct's validity are good (Hair et al. 2006).

The internal reliability of the measurement models was tested using Fornell's composite reliability (Fornell and Larcker 1981). Reliability of the factors was estimated by checking composite reliability. Composite reliability should be greater than the benchmark of 0.7 to be considered adequate (Fornell and Larcker 1981). The formula for calculating composite reliability is as follows:

$$\text{Composite Reliability}\ (\rho) = \frac{(\sum \lambda_i)^2}{\left[(\sum \lambda_i)^2 + (\sum \delta_i)\right]}$$

where λ is the standardized factor loadings and δ is the indicator measurement error.

The composite reliability and AVE'S of all constructs are presented in Table 5.14. All composite reliabilities of constructs have a value higher than 0.70, indicating adequate internal consistency.

5.8.3 *Discriminant Validity*

Discriminant validity is the extent to which a construct is truly distinct from other constructs. It means that a latent variable should explain better the variance of its own indicators than the variance of other latent variables. In other words, the loading of an indicator on its assigned latent variable should be higher than its loadings on all other latent variables.

Discriminant validity check is done by comparing the AVE's with the squared correlation for each of the constructs. The AVE of a latent variable should be higher than the squared correlations between the latent variable and all other latent variables.

Table 5.14 Composite reliability

Construct	CR (composite reliability)	AVE
IT-enabled KM activities	0.755	0.607
Recognition and contribution of best facilities	0.786	0.648
Intranet and Internet platform	0.817	0.598
Communications are facilitated	0.747	0.597
Top Management support	0.756	0.608
Colloquium and seminar	0.855	0.663
KM in actions	0.803	0.671
Uses of knowledge tools	0.745	0.594
Periodical recognition for contribution	0.826	0.614
Learning from experience	0.857	0.750
Collation of knowledge and its capitalization	0.773	0.630

The rule of thumb for assessing discriminant validity requires that the square toot of AVE be larger than the squared correlations between constructs (Hair et al. 2010).

Discriminant validity is shown when each measurement item correlates weakly with all other constructs except for the one to which it is theoretically associated. Discriminant validity is shown when two things happen:

- The correlation of the latent variable score with measurement item need to show an appropriate pattern of loading, one in which the measurement item load highly on their theoretically assigned factor and not highly on other factors.
- Establishing discriminant validity requires an appropriate AVE (average variance extracted) analysis. The test is to see whether the square root of every AVE for each construct is much larger than any correlation among any pair of latent construct. As a rule of thumb, the square root of each construct should be much larger than the correlation of the specific construct with any of the other constructs in the model and should be at least 0.50 (Fornell and Larker 1981).

Discriminant validity is established on the basis of AVE and maximum shared variance (MSV). Criteria for ensuring discriminant validity are MSV < AVE and average shared variance (ASV) < AVE (Hair et al. 2010). Within the present study, MSV and ASV for each of the five individual constructs have been determined. Thus, measurement model was found to be majorly valid in terms of disc ruminant validity as both MSV and ASV of four out of the five individual constructs have been found to be lower than their respective AVE estimates (Table). 'Policy' construct indicated validity concerns with respect to discriminant validity. The discriminant validity statistics for the individual constructs were determined using Microsoft Excel-based Validity Concerns Toolkit developed by Prof. Gakingston.

Shared variance is the amount of variance that a variable (construct) is able to explain in another variable (construct). It is represented by the square of the correlation between any two variables (constructs). For example, if the correlation between two variables, $x1$ and $x2$, is 0.6, then the shared variance between $x1$ and $x2$ is 0.36. If independent variables are correlated, they share some of their predictive power over dependent variables (Hair et al. 2006). Inspection of the correlation matrix between latent constructs can often identify potential shared variance issues.

Discriminant Validity

- MSV < AVE
- ASV < AVE

Square root of AVE greater than inter-construct correlations.

If you have convergent validity issues, then your variables do not correlate well with each other within their parent factor; that is, the latent factor is not well explained by its observed variables. If you have discriminant validity issues, then your variables correlate more highly with variables outside their parent factor than with the variables within their parent factor; that is, the latent factor is better explained by some other variables (from a different factor), than by its own observed variables. Table 5.15 signifies that the discriminant validity has been established. MSV and ASV of each factor shows the proper sharing of variances

Table 5.15 Discriminant validity

Construct	CR	AVE	MSV	ASV
IT-enabled KM activities	0.755	0.607	0.604	0.417
Recognition and contribution of best facilities	0.786	0.648	0.604	0.395
Intranet and Internet platform	0.817	0.598	0.514	0.299
Communications are facilitated	0.747	0.597	0.453	0.369
Top Management support	0.756	0.608	0.356	0.249
Colloquium and seminar	0.855	0.663	0.461	0.362
KM in actions	0.803	0.671	0.453	0.318
Uses of knowledge tools	0.745	0.594	0.386	0.238
Periodical recognition for contribution	0.826	0.614	0.514	0.369
Learning from experience	0.857	0.750	0.486	0.349
Collation of knowledge and its capitalization	0.773	0.630	0.500	0.387

with the AVE. Maximum AVE is explained in learning from experience which is having least error with highest composite reliability. Other factors contain the same property on the basis of the CR, AVE, MSV and ASV scores.

To examine discriminant validity, the shared variances between factors were compared with the average variance extracted (AVE) of the individual factors (Fornell and Larcker 1981). The proof of discriminant validity is presented in Table 5.16. The diagonal items in the table represent the square root of AVE's, which is a measure of variance between construct and its indicators, and the off diagonal items represent squared correlation between constructs. As we have explained that all the factors of KM should be correlated among themselves due to KM implementation process. KM implementation can be done when all the factors support each other. We can identify the strength of the correlation among the variables in the table. The most strongly inter-correlated items are **IT-enabled KM activities** and **recognition and contribution of best facilities of KM**. These two factors are most important, similarly others factors are correlated and interdependent. Collectively these factors support the KM implementation in the organization.

5.8.4 *Nomological Validity*

Nomological validity refers to ascertaining logical relation between a particular model construct and items on which the same is reflected upon. KM model has been developed and considered for the study has been duly checked for nomological validity in terms of construct-item relatedness. Estimate relationships between the spirituality scale and measures of other constructs with which they should, theoretically, be related in a broader nomological network of relationships among constructs and associated measures.

The purpose of the nomological validation is to determine whether constructs reflected by the KM measurement scale related to other constructs as would be

Table 5.16 Result of confirmatory factor analysis on oblique rotation

Construct	IT-enabled KM activities	Recognition and contribution of best facilities	Intranet and Internet platform	Communications are facilitated	Top Management support	Colloquium and seminar	KM in actions	Uses of knowledge tools	Periodical recognition for contribution	Learning from experience	Collation of knowledge and its capitalization
IT-enabled KM activities	0.779										
Recognition and contribution of best facilities	0.777	0.805									
Intranet and Internet platform	0.568	0.544	0.773								
Communications are facilitated	0.671	0.669	0.638	0.773							
Top Management support	0.589	0.508	0.414	0.420	0.780						
Colloquium and seminar	0.679	0.605	0.511	0.573	0.568	0.814					
KM in actions	0.673	0.626	0.464	0.573	0.597	0.652	0.819				
Uses of knowledge tools	0.468	0.422	0.356	0.621	0.498	0.532	0.442	0.771			
Periodical recognition for contribution	0.660	0.653	0.717	0.625	0.503	0.558	0.492	0.535	0.784		
Learning from experience	0.634	0.697	0.586	0.570	0.402	0.655	0.564	0.451	0.587	0.866	
Collation of knowledge and its capitalization	0.693	0.707	0.579	0.673	0.443	0.660	0.504	0.500	0.696	0.693	0.794

predicted on the basis of previous research, e.g. Theories (Mallard and Lance 1998). Niehoff and Moorman (1993, p. 537) examined the ‘nomological network validity’ of a new monitoring measure by correlating it with other leadership measures to demonstrate convergent validity.

In the given model of KM, it would be identified that how KM scales are related with the factors and factors are inter-related with other constructs. Model support that KM implementation in the organization can be implemented effectively when all the given constructs simultaneously work together. This is technologically accepted model by the analysis done in AMOS 20, the model are representing the 11 constructs (factors) along with 25 observed variables.

On the basis of available data and its transformation, fit measurement model is representing 11 factors which are highly correlated and 25 variables supporting the factors. The model can be implemented in the organization by taking care of the defined factors and variables through strategic orientation. Figure 5.7 of model supports the statement along with statistical scores explained earlier.

5.9 Measurement Model and Its Confirmation

After measuring the reliability and validity of the constructs and the variables of the model, the results of confirmatory factor analysis (CFA) using AMOS 20 have been used to evaluate the model fit for the measurement model. Eclipses indicate latent factor, rectangles indicate observed variables and *e* indicates error coefficient in the Fig. 5.7.

5.9.1 Model Identification

Structural models may be just-identified, over-identified or under-identified. A just-identified model is one in which there is a one to one correspondence between the data and the structural parameters. That is, the number of data variances and co variances equals the number of parameters to be estimated. An under-identified model is one which the number of parameters to be estimated exceeds the number of variances and covariances. As such, the model would contain insufficient information for attaining a solution.

Under identification occurs when not enough relevant data are available to obtain unique parameter estimates. Note that when the degrees of freedom of a model are negative, at least one or its parameters is under-identified. However, positive degrees of freedom do not prove that each parameter is identified. ‘Thus, negative degree of freedom is a sufficient but not a necessary criterion for showing that a model is globally under-identified’.

Just-identified models are always identified in a trivial way: Just-identification occurs when the number of data elements equals the number of parameters to be estimated. If the model is just-identified, a solution can always be found for the

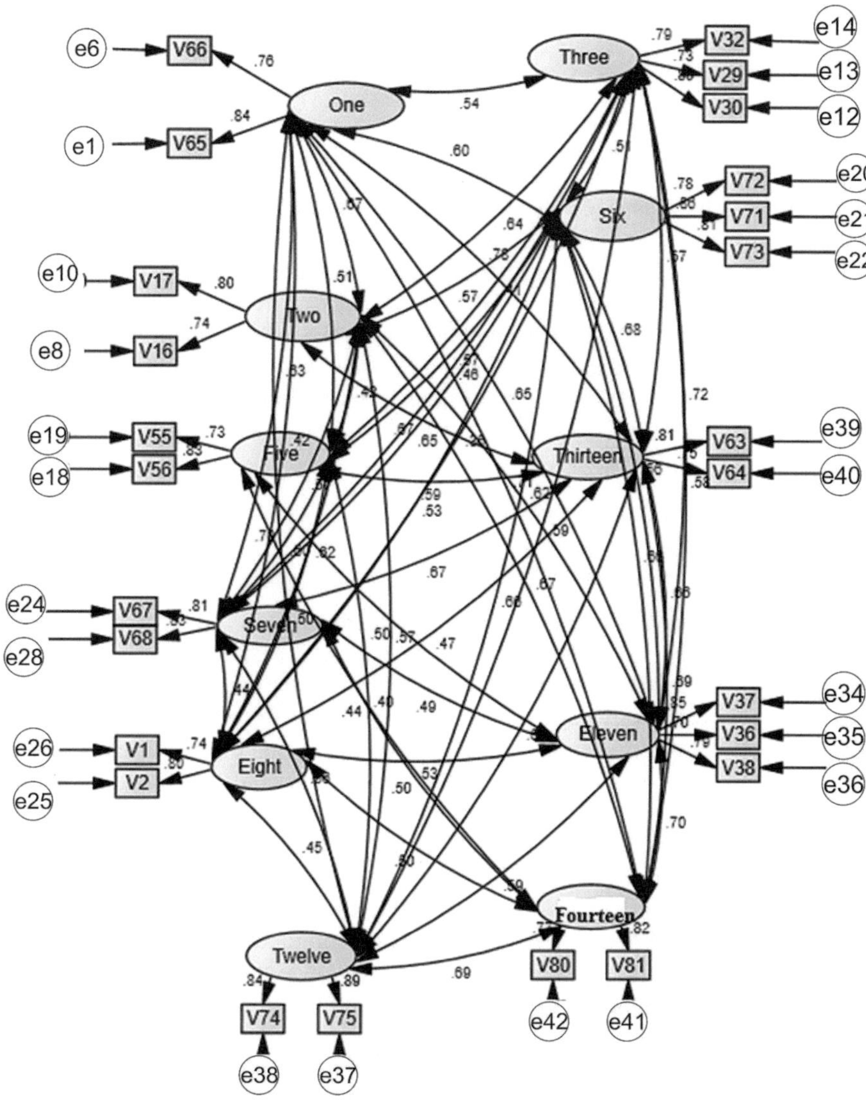

Fig. 5.7 AMOS-based TAM

parameter estimates that will result in perfect fit a discrepancy function equal to zero. This implies both the chi-square goodness-of-fit test and its associated degrees of freedom will be zero. As such, there is no way one can really test/confirm the veracity of a just-identified model. In fact, all just-identified models are equivalent, but not all equivalent models just-identified.

If theory testing is one's objective, the most desirable identification status of a model is over identification, where number of available data elements is more than

Table 5.17 Model detail

Number of distinct sample moments	325
Number of distinct parameters to be estimated	105
Degrees of freedom (325 − 105)	220

what we need to obtain a unique solution. In an over-identified model, degrees of freedom number of data elements greater than the number of parameters to be estimated so that we can explicitly test model fit.

The proposed model in this study is an over-identified model with positive degrees of freedom (220) as shown in Table 5.17 drawn from the AMOS output. In this model, there are 325 distinct sample moments (i.e. pieces of information) from which to compute the estimates of the default model, and 105 distinct parameters to be estimated, leaving 220 degrees of freedom, which is positive (greater than zero). Hence, the model is an over-identified one.

5.9.2 Model Estimation

The most widely used estimation method is maximum likelihood (ML) estimation. The term maximum likelihood describes the statistical principle that underlies the derivation of parameter estimates: the estimates are the ones that maximize the likelihood (the continuous generalization) that the data (the observed co variances) were drawn from this population. That is, ML estimators are those that maximize the likelihood of a sample that is actually observed (Winer et al. 1991). It is a normal theory method because ML estimation assumes that the population distribution for the endogenous variables is multivariate normal. Other methods are based on different parameter estimation theories, but they are not currently used as often. In fact, the use of an estimation method other than ML requires explicit justification (Hoyle 1995). Most forms of ML estimation in SEM are simultaneous, which means that estimates of model parameters are calculated all at once. For this reason, ML estimation is described in the statistical literature as a full-information method.

The method of ML estimation is very complicated and is often iterative, which means that the computer derives an initial solution and then attempts to improve these estimates through subsequent cycles of calculations. 'Improvement' means that the overall fit of the model to the data generally becomes better from step to step. For most just-identified structural equation models, the fit will eventually be perfect. Iterative estimation may converge to a solution quicker if the procedure is given reasonably accurate start values, which are initial estimates of a model's parameters. If these initial estimates are grossly inaccurate—for instance, the start value for a path coefficient is positive when the actual direct effect is negative—then iterative estimation may fail to converge, which means that a stable solution has not been reached. Iterative estimation can also fail if the relative variances among the observed variables are very different; that is, the covariance matrix is ill scaled.

In this study, the minimum iteration was achieved, thereby providing an assurance that the estimation process yielded an admissible solution, eliminating any concern about multicolinearity effects.

5.9.3 Model Evaluation

Goodness of Fit
Ideally evaluation of a model fit should derive from a variety of perspectives and be based on several criteria that assess model fit from a diversity of perspectives.

The model fitting process involves determining the goodness of fit between the hypothesized model and the sample data. Goodness of fit (GOF) indicates how well the specified model reproduces the observed covariance matrix among the indicator items (i.e. the similarity of the observed and estimated covariance matrices). Ever since the first GOF measure was developed, researchers have been striving to refine and develop new measures that reflect various facets of the model's ability to represent the data. As such, a number of alternative GOF measures are available to the researcher. Each GOF measure is unique, but the measures are classified into three general groups: absolute measures, incremental measures and parsimony fit measures.

Once the specified model is estimated, model fit compares the theory to reality by assessing the similarity of the estimated covariance matrix (theory) to reality (the observed covariance matrix). If the theory is perfect, the observed and estimated covariance matrices would be the same. The values of any GOF measure result from a mathematical comparison of these two matrices. The closer the values of these two matrices are to each other, the better the model is said to fit. Given below is a description of the goodness-of-fit indicators used to evaluate model fitness in structural equation modelling (SEM).

5.9.3.1 Chi-Squared (χ^2) Goodness of FIT

The chi-squared goodness of fit metric is used to assess the correspondence between theoretical specification and empirical data in a CFA. By default, the null hypothesis of SEM is that the observed sample and SEM-estimated covariance matrices are equal, meaning perfect fit. The chi-square value increases as differences (residuals) are found when comparing the two matrices. With the chi-square test, the statistical probability that the observed sample and SEM-estimated covariance matrices are equal is assessed. The probability is the traditional *p*-value associated with parametric statistical tests.

This statistic (χ^2) is also known as the likelihood ratio chi-square or generalized likelihood ratio. The estimation process in SEM will focus on yielding parameter values so that the discrepancy between sample covariance matrix (S) and the SEM-estimated covariance matrix is minimal. The value χ^2 of for a just-identified model generally equals zero and has no degrees of freedom. If $\chi^2 = 0$, the model perfectly fits the data (i.e. the predicted correlations and covariance's equal their observed counterparts). As the value of chi-square increases, the fit of an over-identified model becomes increasingly worse. Thus, chi-square is actually a 'badness-of-fit' index because the higher its value, the worse the model's correspondence to the data.

5.9.3.2 Goodness-of-Fit Index (GFI and AGFI)

The goodness-of-fit index (GFI) was the very first standardized fit index (Joreskog and Sorbom 1981). It is analogous to a squared multiple correlation () except that the GFI is a kind of matrix proportion of explained variance. Thus, GFI = 1.0 indicates perfect model fit, GFI > 0.90 may indicate good fit, and values close to zero indicate very poor fit. However, values of the GFI can fall outside the range 0–1.0. Values greater than 1.0 can be found with just-identified models or with over-identified models with almost perfect fit; negative values are most likely to happen when the sample size is small or when model fit is extremely poor.

Another index originally associated with AMOS is the adjusted goodness-of-fit index (AGFI; Joreskog and Sorbom 1981). The AGFI differs from the GFI only in the fact that it adjusts for the number of degrees of freedom in the specified model. The GFI and AGFI can be classified as absolute indices. The parsimony goodness-of-fit index (PGFI; Mulaik et al. 1989) corrects the value of the GFI by a factor that reflects model complexity, but it is sensitive to model size.

5.9.3.3 Normed Fit Index

The NFI is one of the original incremental fit indices introduced by Bentler and Bonnet (1980). It is a ratio of the difference in the χ^2 value for the fitted model and the null model divided by the χ^2 value for the null model. It ranges between zero to one. A normed fit index of one indicates perfect fit.

5.9.3.4 Relative Fit Index

The relative fit index (RFI; Bollen and Long 1993) represents a derivative of the NFI; as with both the NFI and CFI, the RFI coefficient values range from zero to one with values close to one indicating superior fit (Hu and Bentler 1999).

5.9.3.5 Comparative Fit Index

The CFI is an incremental fit index that is an improved version of the NFI (Bentler and Bonnet 1980; Hu and Bentler 1999). The CFI is normed so that values range between 0 and 1, with higher values indicating better fit. Because the CFI has many desirable properties, including its relative, but not complete, insensitivity to model complexity, it is among the widely used indices. CFI values above 0.90 are usually associated with a model that fits well. But a revised cut off value close to 0.95 was suggested by Hu and Bentler (1999).

5.9.3.6 Tucker Lewis Index

The Tucker–Lewis index (Tucker and Lewis 1973) is conceptually similar to the NFI, but varies in that it is actually a comparison of the normed chi-square values for the null and specified model, which to some degree takes into account model complexity. Models with good fit have values that approach one (Hu and Bentler 1999), and a model with a higher value suggests a better fit than a model with a lower value.

5.9.3.7 Root Mean Square Error of Approximation (RMSEA)

Root mean square error approximation (RMSEA) was first proposed by Steiger and Lind (1980). It is one of the most widely used measures that attempts to correct for the tendency of the χ^2 GOF test statistic to reject models with a large sample or a large number of observed variables. Thus, it better represents how well a model fits a population, not just the sample used for estimation. Lower RMSEA values indicate better fit. Earlier research suggests values of <0.05. Browne and Cudeck (1993), Hu and Bentler (1999) have suggested value of <0.06 to be indicative of good fit.

5.9.3.8 Validating the Measurement Model

The results shown in Table 5.18 provide a quick overview of the model fit, which includes the value (464.407), together with its degrees of freedom (220) and probability value (0.000).

In the table, NPAR stands for number of parameters, and CMIN (χ^2) is the minimum discrepancy and represents the discrepancy between the unrestricted sample covariance matrix *S* and the restricted covariance matrix. df stands for degrees of freedom, and *P* is the probability value.

Table 5.18 Model fit table of AMOS output

Model	NPAR	CMIN	DF	*P*	CMIN/DF
Default model	105	464.407	220	0.000	2.111
Saturated model	325	0.000	0		
Independence model	25	5740.989	300	0.000	19.137

Table 5.19 Fit statistics of the model

Fit statistic	Recommended	Obtained
χ^2	…	467.407
df	…	220
χ^2 significance	$p \leq 0.05$	0
χ^2/df	<5.0	2.111
GFI	>0.90	0.919
AGFI	>0.85	0.88
NFI	>0.90	0.919
RFI	>0.90	0.9
CFI	>0.95	0.955
TLI	>0.90	0.939
RMSEA	≤0.05	0.5

Although the χ^2 found good, it is also appropriate to check the value of χ^2 divided by df (Wheaton et al. 1977) as the statistic is particularly sensitive to sample sizes (i.e. the probability of model rejection increases with increasing sample size, even if the model is minimally false), and hence, chi-square (χ^2) divided by degrees of freedom is suggested as a better fit metric (Bentler and Bonnett 1980). For the current CFA model, as shown in Table 5.19, χ^2/df was 2.111 (=464.407; df = 220), suggesting acceptable model fit.

Goodness-of-fit index (GFI) obtained is 0.92 as against the recommended value of above 0.90. The adjusted goodness-of-fit index (AGFI) is 0.88 as against the recommended value of above 0.85 as well. The normed fit index (NFI), relative fit index (RFI), comparative fit index (CFI), Tucker–Lewis index (TLI) are 0.92, 0.90, 0.95, 0.94, respectively, as against the recommended level of above 0.90. RMSEA is 0.05 and is equal to the recommended limit of 0.05. Hence, the model shows an overall acceptable fit. The model is an over-identified model.

The confirmatory factor analysis shows that how the model can be proved. The constructs and variables verified all the level of goodness of fit. Model explains that how KM constructs are inter-correlated and need to be implements collectively. Implementation of information technology and contribution of best facilities are showing the strong relation means together these two strengthen the KM implementation.

5.10 Hypotheses Testing

Friedman's Test (Nonparametric Test Algorithms)

Eighty one variables have been constructed on the basis of twelve well-defined groups. Each group is strongly related with the respective variables. On the basis of 12 well-defined groups, related variables are divided and tested.

Each group belongs to same well-defined subject, which is measured with Friedman test (Table 5.20).

Table 5.20 Grouping the variables

S. No.	Group variables	Group name
1	11. Knowledge is treated as an asset	Knowledge is an asset
	12. Knowledge is treated as an object	
	13. Knowledge is used as an important and critical resource in my dept.	
2	7. Appropriate values and mindset towards KM	Attitude and behaviour
	8. Personnel in POWERGRID are oriented towards KM	
	69. Employees in the department are willing to share the knowledge	
3	2. Knowledge mgmt. in activities	Culture
	9. Aesthetically oriented culture for KM is the way of life	
	10. Open-minded and collaborative working	
	17. Innovation and knowledge creation are way of life in POWERGRID	
	18. Reward and recognition for suggestion of best practices have been incorporated	
	23. Personal best practices can be shared with other employee easily	
	28. Continuously strives for knowledge acquisition, dissemination and capture from its employee	
	39. Collegial, flat, team-based, open-door Policy	
	49. Programs have periodic team meetings	
	51. Team/department employee share their knowledge in informal manner	
	55. Communications are sent through circular or memos in electronic environment	
	59. Contact Communities of Practice in my area and member of these	
	61. Acquisition of knowledge is a regular practice	
	62. Knowledge is a way of life in every employees' job	
	66. Mandatory for every employee to documents insight	
	67. Sponsor trips to international/national conferences and seminars	
	68. Arrange seminar lecture by external experts	
	70. Free access of information	
	76. Actively promote 'culture of experimentation' that encourages employees to try new ideas and take risk	
4	2. Knowledge mgmt. in activities	Strategy (For Knowledge Management)
	3. Subject-matter experts available in the department	
	4. Same mistake done by someone in the past is being repeated	
	5. Lots of documents/reports lying unutilized in my dept.	
	6. Time-consuming to access any documents in my department	
	14. Vision in KM which properly articulates its knowledge needs	
	15. KM is being used constantly as a business strategy	
	16. Transfer of knowledge and best practices are taking place in day-to-day working of employee	
	24. Cohesive and powerful vision of the organization is shared across the workforce to promote strategic thinking	
	25. Enabling structure in terms of hierarchy and communication flow	
	65. Employee is encouraged to periodically produce documents	
	77. Mission, vision and objective of KM clearly propagated throughout	
	78. ERP provides further boosts to KM	

(continued)

Table 5.20 (continued)

S. No.	Group variables	Group name
5	25. Enabling structure in terms of hierarchy and communication flow	Team-based matrix structure
	26. Cohesive teems which facilitate sharing of experience and information	
	35. 'Chief knowledge officer' for managing the knowledge-related activities	
	40. Matrix form of structure in the Company	
	41. Brick-and-mortar model	
6	1. Knowledge mgmt. in POWERGRID	Well-laid system
	19. Information about the outcome of implementing best practices is translated in practice in the dept.	
	20. Dedicated staff for knowledge capture and acquisition are earmarked in the department	
	21. Information regarding relevant topics is readily available	
	47. Meeting/conference rooms are available for team meeting and dissections	
	48. Employees/team can discuss work through mobile, teleconferencing, VOIP, etc.	
	49. Programs have periodic team meetings	
	50. Every programme has closure meeting to critically review the programme	
	57. Directory of personnel contains information of names of people	
	64. Capitalization of knowledge	
	66. Mandatory for every employee to documents insight	
	69. Employees in the dept. are willing to share the knowledge	
7	23. Personal best practices can be shared with other employee easily	Predefined process (KM)
	79. Organization measures and manages its intellectual capital in a systematic way	
8	71. Support from mgmt. for KM	Management support
	72. Superiors fully endorse way of handling KM-related issues	
	73. Top mgmt. supports and encourages KM	
9	42. Employee is imparted training required by him for his job	Relevant training
	43. KM training is identified as part of organizational training need assessment	
10	18. Reward and recognition for suggestion of best practices have been incorporated	Reward and recognition
	74. POWERGRID periodically gives special award to employees for KM	
	75. Recognition in the form of titles, certificate and commendation letter is periodically awarded	
	76. Actively promotes 'culture of experimentation' that encourages employees to try new ideas and take risks	

(continued)

Table 5.20 (continued)

S. No.	Group variables	Group name
11	37. Knowledge maps and matrices of the Company used for knowledge requirement	Tools (related with HR)
	38. Contacting subject-matter expert by using corporate yellow pages	
	44. Organization has central repository/library	
	45. Documents of the organization are available in electronic formats	
	46. Documents are arranged in such a manner so that it can be easily accessed	
	53. Newsletter/bulletin will inform about the happening	
	56. Noticeboards are kept on vantage points	
	57. Directory of personnel contain information of names of people	
	58. nformation about the location of project report, research facilities and equipment available is displayed prominently	
	60. Knowledge is created in technology development department and in substations	
	61. Acquisition of knowledge is a regular practice	
	63. Knowledge Management Cell collates knowledge nuggets and disseminates it through power portal	
	80. POWERGRID publishes regular intellectual capital report	
12	29. Employees are provided 'instant messenger' or 'chat'—videoconferencing facilities	Information and communication
	30. Online discussion through intranet	
	31. Expert database	
	32. Blogging platform	
	33. Employees upload their documents on the power portal	
	34. Download any documents from power portal	
	36. Visit knowledge power portal for needs	
	44. Organization has central repository/library	
	46. Documents are arranged in such a manner so that it can be easily accessed	
	48. Employees/team can discuss work through mobile, teleconferencing, VOIP, etc.	
	54. Intranet/Website plays a vital role in disseminating information to the employees	
	55. Communications are sent through circular or memos in electronic environment	
	56. Noticeboards are kept on vantage points	
	58. Information about the location of project report, research facilities and equipment available is displayed prominently	
	63. Knowledge Management Cell collates knowledge nuggets and disseminates it through power portal	
	65. Employee is encouraged to periodically produce documents	
	70. Free access to information	
	78. ERP provides further boosts to KM	
	80. POWERGRID publishes regular intellectual capital report	

Sum of Ranks

The scores in each column are ranked, r_{ij} denoting the rank within block j of the observation in treatment ji. Average ranks are assigned to tie scores.

1. The ranks are summed over each treatment to give rank sums, for

$$t_i = \sum_{j=1}^{n} r_{ij} \quad \text{for } i = 1, 2, \ldots, k$$

2. Friedman test statistic FR is calculated as

$$\text{FR} = \frac{12}{nk(k+1)} \sum_{i=1}^{n} \left\{ t_i - \frac{1}{2} n(k+1) \right\}^2$$

The significance level is compared to the x^2 distribution with $k - 1$ degrees of freedom, where k is the total number of samples

k number of variables
n number of cases
χ^2 Chi-square = FR

The significance level is from the χ^2 distribution with $k - 1$ degrees of freedom.

Mean Rank Let $x1$, $x2$, …, xn be the n samples in an event series. The samples are ranked in increasing order and the rank r_i for each sample is calculated. In case of ties, we assign average rank to each sample.

H_1 Knowledge as an asset can be managed and
H_0 Knowledge as an asset cannot be managed.

Friedman Test

Table shows the mean rank of the variables of H_1 (Tables 5.21 and 5.22).

Table shows the SPSS output of the test statistics. The value of this statistics is 38.473 and its degree of freedom ($K - 1$) is $3 - 1 = 2$. The significance value is 0.000, which is less than 0.05 (at 95 % of confident level); therefore, we can conclude that null hypothesis has been rejected. So, knowledge is an asset and has significant role in management.

H_2 the attitude and behaviour of personnel help Knowledge Management Implementation positively and

Table 5.21 Friedman test

Ranks	
	Mean rank
11. Knowledge is treated as an asset	2.09
12. Knowledge is treated as an object	1.84
13. Knowledge is used as an important and critical resource in my dept.	2.07

H_0 the attitude and behaviour of personnel do not help Knowledge Management Implementation positively.

Table shows the mean rank of the inter-related variables of H_2 (Tables 5.23 and 5.24).

Table shows the SPSS output of the test statistics. The value of this statistics is 14.243 and its degree of freedom ($K - 1$) is $3 - 1 = 2$. The significance value is 0.001, which is less than 0.05 (at 95 % of confident level); therefore, we can conclude that null hypothesis has been rejected. So, the attitude and behaviour of personnel play significant role in implementation of Knowledge Management in POWERGRID.

H_3 Culture is the prime mover of Knowledge Management Implementation and
H_0 Culture is not the prime mover of Knowledge Management Implementation.

Table shows the mean rank of the inter-related variables of H_3 (Tables 5.25 and 5.26).

Table shows the SPSS output of the test statistics. The value of this statistics is 38.473 and its degree of freedom ($K - 1$) is $3 - 1 = 2$. The significance value is 0.000, which is less than 0.05 (at 95 % of confident level); therefore, we can

Table 5.22 Friedman test

Test statistics[a]	
N	414
Chi-square	38.473
df	2
Asymp. sig.	0.000
Exact sig.	0.000
Point probability	0.000

[a]Friedman test

Table 5.23 Friedman test

Ranks	
	Mean rank
7. Appropriate values and mindset towards KM	1.91
8. Personnel in POWERGRID are oriented towards KM	1.98
69. Employee in the department are willing to share the knowledge	2.10

Table 5.24 Friedman test

Test statistics[a]	
N	414
Chi-square	14.243
df	2
Asymp. sig.	0.001

[a]Friedman test

Table 5.25 Friedman test

Ranks	
	Mean rank
2. Knowledge mgmt. in activities	10.73
9. Aesthetically oriented culture for KM is the way of life	9.68
10. Open-minded and collaborative working	10.74
17. Innovation and knowledge creation are way of life in PGCIL	9.72
18. Reward and recognition for suggestion of best practices have been incorporated	8.72
23. Personal best practices can be shared with other employee easily	10.26
28. Continuously strives for knowledge acquisition, dissemination and capture from its employee	10.21
39. Collegial, flat, team-based, open-door Policy	8.18
49. Programs have periodic team meetings	11.14
51. Team/department employee share their knowledge in informal manner	11.23
55. Communications are sent through circular or memos in electronic environment	11.93
59. Contact Communities of Practice in my area and member of these	7.69
61. Acquisition of knowledge is a regular practice	11.31
62. Knowledge is a way of life in every employees' job	12.33
66. Mandatory for every employee to document insight	6.96
67. Sponsor trips to international/national conferences and seminars	9.93
68. Arrange seminar lecture by external experts	11.57
70. Free access to information	9.39
76. Actively promote 'culture of experimentation' that encourages employees to try new ideas and take risks	8.28

Table 5.26 Friedman test

Test statistics[a]	
N	414
Chi-square	741.175
df	18
Asymp. sig.	0.000

[a]Friedman test

conclude that null hypothesis has been rejected. So, knowledge is an asset and has significance role in the management of Company.

H_4 Strategy impacts Knowledge Management Implementation directly and
H_0 Strategy does not impact Knowledge Management Implementation directly.

Table shows the mean rank of the inter-related variables H_4 (Tables 5.27 and 5.28).

Table shows the SPSS output of the test statistics. The value of this statistics is 444.879 and its degree of freedom $(K - 1)$ is $13 - 1 = 12$. The significance value is

Table 5.27 Friedman test

Ranks	
	Mean rank
2. Knowledge mgmt. in activities	7.85
3. Subject-matter experts available in the department	6.19
4. Same mistake done by someone in the past is being repeated	6.03
5. Lots of documents/reports lying unutilized in my dept.	5.38
6. Time-consuming to access any documents in my department	5.67
14. Vision in KM properly articulates its knowledge needs	7.78
15. KM is being used constantly as a business strategy	7.32
16. Transfer of knowledge and best practices are taking place in day-to-day working of employee	7.27
24. Cohesive and powerful vision of the organization is shared across the workforce to promote strategic thinking	7.48
25. Enabling structure in terms of hierarchy and communication flow	7.01
65. Employee is encouraged to periodically produce documents	6.40
77. Mission, vision and objective of KM clearly propagated throughout Company	7.57
78. ERP provides further boosts to KM	9.05

Table 5.28 Friedman test

Test statistics[a]	
N	414
Chi-square	444.879
df	12
Asymp. sig.	0.000

[a]Friedman test

0.000, which is less than 0.05 (at 95 % of confident level); therefore, we can conclude that null hypothesis has been rejected. So, strategy impacts Knowledge Management significantly.

H_5 Team-based matrix structure affects Knowledge Management Implementation and

H_0 Team-based matrix structure does not affect Knowledge Management Implementation.

Table shows the mean rank of the inter-related variables of H_5 (Tables 5.29 and 5.30).

Table shows the SPSS output of the test statistics. The value of this statistics is 62.607 and its degree of freedom ($K - 1$) is $5 - 1 = 4$. The significance value is 0.000, which is less than 0.05 (at 95 % of confident level); therefore, we can conclude that null hypothesis has been rejected. So, team-based matrix structure significantly affects Knowledge Management Implementation.

Table 5.29 Friedman test

Ranks	
	Mean rank
25. Enabling structure in terms of hierarchy and communication flow	3.21
26. Cohesive terms which facilitates sharing of experience and information	3.15
35. 'Chief knowledge officer' for managing the knowledge-related activities	2.60
40. Matrix form of structure in the Company	3.01
41. Brick-and-mortar model	3.03

Table 5.30 Friedman test

Test statistics[a]	
N	414
Chi-square	62.607
df	4
Asymp. sig.	0.000

[a]Friedman test

H_6 Well-laid systems enable Knowledge Management Implementation suitably and
H_0 Well-laid systems do not enable Knowledge Management Implementation suitably.

Table shows the mean rank of the inter-related variables of H_6 (Tables 5.31 and 5.32).

Table shows the SPSS output of the test statistics. The value of this statistics is 827.093 and its degree of freedom $(K - 1)$ is $12 - 1 = 11$. The significance value is 0.000, which is less than 0.05 (at 95 % of confident level); therefore we can

Table 5.31 Friedman test

Ranks	
	Mean Rank
1. Knowledge mgmt. in POWERGRID has taken place	8.01
19. Information about the outcome of implementing best practices is translated in practice in the dept.	5.79
20. Dedicated staff for knowledge capture and acquisition are earmarked in the department	4.58
21. Information regarding relevant topics are readily available	6.14
47. Meeting/conference rooms are available for team meeting and dissections	7.21
48. Employees/team can discuss work through mobile, teleconferencing, VOIP, etc.	7.79
49. Programs have periodic team meetings	6.92
50. Every programme has closure meeting to critically review the programme	6.08
57. Directory of personnel contains information of names of people	8.59
64. Capitalization of knowledge done through consultancy	5.69
66. Mandatory for every employee to documents insight	4.37
69. Employees in the dept. are willing to share the knowledge	6.83

Table 5.32 Friedman test

Test statistics[a]	
N	414
Chi-square	827.093
df	11
Asymp. sig.	0.000

[a]Friedman test

conclude that null hypothesis has been rejected. So, well-laid systems make significant impact on enabling Knowledge Management suitably.

H_7 Predefined processes help Knowledge Management Implementation and
H_0 Predefined processes do not help Knowledge Management Implementation.

Table shows the mean rank of the inter-related variables H_7 (Tables 5.33 and 5.34).

Table shows the SPSS output of the test statistics. The value of this statistics is 33.640 and its degree of freedom ($K - 1$) is $2 - 1 = 1$. The significance value is 0.000, which is less than 0.05 (at 95 % of confident level), therefore we can conclude that, null hypothesis has been rejected. So, predefined processes make significant role in Knowledge Management Implementation.

H_8 Management support is prerequisite for Knowledge Management and
H_0 Management support is not prerequisite for Knowledge Management.

Table shows the mean rank of the inter-related variables H_8 (Tables 5.35 and 5.36).

Table shows the SPSS output of the test statistics. The value of this statistics is 11.170 and its degree of freedom ($K - 1$) is $3 - 1 = 2$. The significance value is

Table 5.33 Friedman test

Ranks	
	Mean rank
23. Personal best practices can be shared with other employee easily	1.61
79. Organization measures and manages its intellectual; capital in a systematic way	1.39

Table 5.34 Friedman test

Test statistics[a]	
N	414
Chi-square	33.640
df	1
Asymp. sig.	0.000

[a]Friedman test

Table 5.35 Friedman test

Ranks	
	Mean rank
71. Support from mgmt. for KM	1.94
72. Superior fully endorse way of handling KM-related issues	1.98
73. Top mgmt. supports and encourages KM	2.08

Table 5.36 Friedman test

Test statistics[a]	
N	414
Chi-square	11.170
df	2
Asymp. sig.	0.004

[a]Friedman test

0.004, which is less than 0.05 (at 95 % of confident level); therefore, we can conclude that null hypothesis has been rejected. So, management support is valuable and is a prerequisite for Knowledge Management.

H_9 Relevant training to personnel is required for Knowledge Management Implementation and

H_0 Relevant training to personnel is not required for Knowledge Management Implementation.

Table shows the mean rank of the inter-related variables H_3 (Tables 5.37).

Table shows the SPSS output of the test statistics. The value of this statistics is 5.769 and its degree of freedom $(K - 1)$ is $2 - 1 = 1$. The significance value is 0.016, which is less than 0.05 (at 95 % of confident level); therefore, we can conclude that null hypothesis has been rejected. So, relevant training has significant role in Knowledge Management Implementation.

Table 5.37 Friedman test

Ranks	
	Mean rank
42. Employee is imparted training required by him for his job	1.54
43. KM training is identified as part of organizational training need assessment	1.46
Test statistics[a]	
N	414
Chi-square	5.769
df	1
Asymp. sig.	0.016

[a]Friedman test

H_{10} Reward and recognition are motivators for effective Knowledge Management and

H_0 Reward and recognition are not motivators for effective Knowledge Management.

Table shows the mean rank of the inter-related variables H_{10} (Table 5.38 and 5.39).

Table shows the SPSS output of the test statistics. The value of this statistics is 21.307 and its degree of freedom $(K - 1)$ is $4 - 1 = 3$. The significance value is 0.000, which is less than 0.05 (at 95 % of confident level); therefore, we can conclude that null hypothesis has been rejected. So, reward and recognition play significant role of motivators in Knowledge Management Implementation.

H_{11} Information and communication technology is an enabler for Knowledge Management and

H_0 Information and communication technology is not an enabler for Knowledge Management.

Table shows the mean rank of the inter-related variables of H_{11} (Tables 5.40 and 5.41).

Table shows SPSS output shows of the test statistics. The value of this statistics is 1270.699 and its degree of freedom $(K - 1)$ is $19 - 1 = 18$. The significance value is 0.000, which is less than 0.05 (at 95 % of confident level); therefore, we can conclude that null hypothesis has been rejected. So, information and communication technology is significant as an enabler for Knowledge Management Implementation.

Table 5.38 Friedman test

Ranks	
	Mean rank
18. Reward and recognition for suggestion of Best practices have been incorporated	2.64
74. Periodically give special award to employee	2.43
75. Recognition in form of titles, certificate and commendation letter are periodically awarded	2.37
76. Actively promote 'culture of experimentation' that encourages employees to try new ideas and take risks	2.57

Table 5.39 Friedman test

Test statistics[a]	
N	414
Chi-square	21.307
df	3
Asymp. sig.	0.000

[a]Friedman test

Table 5.40 Friedman test

Ranks	
	Mean rank
29. Employees are provided 'instant messenger' or 'chat'—videoconferencing facilities	6.63
30. Online discussion takes place through intranet	8.07
31. Expert database	8.17
32. Blogging platform	6.40
33. Upload their documents on the power portal	10.49
34. Download any documents from portal	11.76
36. Visit power portal for my knowledge needs	10.20
44. Organization has central repository/library	10.76
46. Documents are arranged in such a manner so that it can be easily accessed	8.34
48. Employees/team can discuss work through mobile, teleconferencing, VOIP, etc.	12.49
54. Intranet/Website plays a vital role in disseminating information to the employee	13.12
55. Communications are sent through circular or memos in electronic environment	11.99
56. Noticeboards are kept on vantage points	11.94
58. Information about location of project report, research facilities and equipment available is displayed prominently	9.58
63. Knowledge Management Cell collates knowledge nuggets and disseminates it through power portal	9.78
65. Employee is encouraged to periodically produce documents	8.66
70. Free access to information	9.51
78. ERP provides further boost to KM	12.61
80. POWERGRID publishes regular intellectual capital report	9.49

Table 5.41 Friedman test

Test statistics[a]	
N	414
Chi-square	1270.699
df	18
Asymp. sig.	0.000

[a]Friedman test

H_{12} Tools related with HR management are helpful for effective Knowledge Management and

H_0 Tools related with HR management are not helpful for effective Knowledge Management.

Table shows the mean rank of the inter-related variables of H_{12} (Table 5.42 and 5.43).

Table shows the SPSS output of the test statistics. The value of this statistics is 708.493 and its degree of freedom $(K - 1)$ is $13 - 1 = 12$. The significance value is

Table 5.42 Friedman test

Ranks	
	Mean rank
37. Knowledge maps and matrices of the Company are used for knowledge requirement	5.91
38. Subject-matter expert may be approached by using my corporate yellow pages	4.99
44. Organization has central repository/library	7.40
45. Documents of the organization are available in electronic formats	6.75
46. Documents are arranged in such a manner so that it can be easily accessed	5.67
53. Newsletter/bulletin will inform about the happening	8.44
56. Noticeboards are kept on vantage points	8.21
57. Directory of personnel contain information of names of people	9.48
58. Information about location of project report, research facilities and equipment available is displayed prominently	6.53
60. Knowledge is created in tech development department and in substations	6.63
61. Acquisition of knowledge is a regular practice	7.79
63. Knowledge Management Cell collates knowledge nuggets and disseminates it through power portal	6.74
80. POWERGRID publishes regular intellectual capital report	6.46

Table 5.43 Friedman test

Test statistics[a]	
N	414
Chi-square	708.493
df	12
Asymp. sig.	0.000

[a]Friedman test

0.000, which is less than 0.05 (at 95 % of confident level); therefore, we can conclude that null hypothesis has been rejected. So, tools related with Knowledge Management are significantly helpful for effective Knowledge Management.

Table reflects the consolidated test results of all the null hypotheses (Table 5.44).

5.11 Integrating Qualitative and Quantitative Data Analysis

This section integrates information from the qualitative and quantitative presentations of the data to draw inferences from the data set. This synthesis is one of the interpretative inferences relying on both deductive and inductive logic, using the qualitative information to determine what was actually said and what it meant, and, the quantitative information to determine where to look. In the foregoing pages, we find that Focus Group discussions with Nodal Officers of Knowledge Management of POWERGRID, interview of Knowledge Management Implementation Committee of POWERGRID and interview of Shri Ravi P. Singh, Director (Personnel),

Table 5.44 Test results of hypotheses

	Hypothesis (H_0, null hypothesis)	Tests	Results
H_1	Knowledge as an asset cannot be managed	Related sample test (Friedman test)	Rejected (null hypothesis)
H_2	The attitude and behaviour of personnel do not help Knowledge Management Implementation positively	Related sample test (Friedman test)	Rejected (null hypothesis)
H_3	Culture is not the prime mover of Knowledge Management Implementation	Related sample test (Friedman test)	Rejected (null hypothesis)
H_4	Strategy does not impact Knowledge Management Implementation directly	Related sample test (Friedman test)	Rejected (null hypothesis)
H_5	Team-based matrix structure does not affect Knowledge Management Implementation	Related sample test (Friedman test)	Rejected (null hypothesis)
H_6	Well-laid systems do not enable Knowledge Management Implementation suitably	Related sample test (Friedman test)	Rejected (null hypothesis)
H_7	Predefined processes do not help Knowledge Management Implementation	Related sample test (Friedman test)	Rejected (null hypothesis)
H_8	Management support is not prerequisite for Knowledge Management	Related sample test (Friedman test)	Rejected (null hypothesis)
H_9	Relevant training to personnel is not required for Knowledge Management Implementation	Related sample test (Friedman test)	Rejected (null hypothesis)
H_{10}	Reward and recognition are not motivators for effective Knowledge Management	Related sample test (Friedman test)	Rejected (null hypothesis)
H_{11}	Information and communication technology is not an enabler for Knowledge Management	Related sample test (Friedman test)	Rejected (null hypothesis)
H_{12}	Tools related with HR management are not helpful for effective Knowledge Management	Related sample test (Friedman test)	Rejected (null hypothesis)

POWERGRID, indicate that management has got proper mechanism in place and has succeeded in implementing Knowledge Management in this case, i.e., in POWERGRID. All the quantitative results make significant impact of Knowledge Management in POWERGRID. All the 15 factors support the predefined research model in totality. Both qualitative and quantitative analyses and their outcome are in perfect sync. The quantitative analysis provides a cueing mechanism, indicating where to direct attention for focused qualitative analysis. The interpretation, however, is ultimately subjective and relies upon the judgement, in this case, of the researcher.

However, confirmatory model, which has been done through structural equation modelling (SEM), contained only eleven constructs. It means model is acceptable on eleven factors only on the basis of the analysis done through SEM. POWERGRID can implement all the constructs collectively those are shown in the Technological Accepted Model (TAM).

References

Albaum, G.: The Likert scale revisited: an alternate version. J. Market Res. Soc. **39**, 331–349 (1997)

Andy, F.: Discovering Statistics using SPSS, 3rd edn. Sage Publication, Washington DC (2009)

Bentler, P.M., Bonnet, D.C.: Significance tests and goodness of fit in the analysis of covariance structures. Psychol. Bull. **88**(3), 588–606 (1980)

Bollen, K.A., Long, J.S.: Testing Structural Equation Models. Sage, Newbury Park, CA (1993)

Brown, M.W., Cudeck, R.: Alternative ways of assessing model fit. In: Bollen, K.A., Long, J.S. (eds.) Testing Structural Equation Models. Sage, Newbury Park, CA (1993)

Carmines, E.G., Zeller, R.A.: Reliability and Validity Assessment. Sage, Beverly Hills, CA (1979)

Coombs, C.H.: A theory of data. Psychol. Rev. **67**, 143–159 (1960)

Diamantopoulos, A., Siguaw, J., Siguaw, J.A.: Introducing LISREL: A Guide for the Uninitiated. Sage Publications, London (2000)

Fornell, C., Larcker, D.F.: Evaluating structural equation models with unobservable variables and measurement error. J. Marketing Res. **18**(1), 39–50 (1981). doi:10.2307/3151312

Garson, G.D.: Structural equation modeling. http://faculty.chass.ncsu.edu/garson/PA765/structur.htm. Accessed 10 May 2011

Hair, Jr. J.F., Black, W.C., Babin, B.J., Anderson, R.E., Tatham, R.L.: Multivariate Data Analysis, 6th edn. Pearson-Prentice Hall, Upper Saddle River, NJ (2006)

Hair, J.F., Black, W.C., Balin, B.J., Anderson, R.E.: Multivariate data analysis. Maxwell Macmillan International Editions, New York (2010)

Hodge, D.R., Gillespie, D.: Phrase completions: an alternative to Likert scales. Soc. Work Res. **27**, 45–55 (2003)

Hoyle, R.H. (ed.): Structural Equation Modeling: Concepts, Issues, and Applications. Sage Publications, Thousand Oaks, CA (1995)

Hu, L.T., Bentler, P.M.: Cutoff criteria for fit indexes in covariance structure analysis: conventional criteria versus new alternatives. Struct. Eqn. Model. **6**(1), 1–55 (1999)

Hutcheson, G.D., Sofroniou, N.: The multivariate social scientist. In: Factor Analysis, pp. 224–225. Sage, London (1999)

Jakobsson, U.: Statistical presentation and analysis of ordinal data in nursing research. Scand. J. Caring Sci. **18**, 437–440 (2004)

Jamieson, S.: Likert scales: how to (ab)use them. Med. Educ. **38**, 1212–1218 (2004)

João, M.: Google scholar accessed on 16th November by using structural equation modeling for ordinal data. Board of Directors of IAVE, I.P. and Associate Professor of Statistics at ISPA-IU (2014)

Joreskog, K.G., Sorbom, D.: LISREL User's Guide. National Educational Services, Chicago (1981)

Knapp, T.R.: Treating ordinal scales as interval scales: an attempt to resolve the controversy. Nurs. Res. **39**, 121–123 (1990)

Kuzon Jr, W.M., Urbanchek, M.G., McCabe, S.: The seven deadly sins of statistical analysis. Ann. Plast. Surg. **37**, 265–272 (1996)

Mallard, A.G.C., Lance, C.E.: Development and evaluation of a parent–employee interrole conflict scale. Soc. Ind. Res. **45**, 537 (1998)

Michael, N.: Google scholar accessed on 16th November by using structural equation modeling for ordinal data Professor at Virginia Commonwealth University) (2014)
Mulaik, S.A., James, L.R., Van Alstine, J., Bennett, N., Lind, S., Stilwell, C.D.: Evaluation of goodness-of-fit indices for structural equation models. Psycho-logical Bull. **105**, 430–445 (1989)
Niehoff, B.P., Moorman, R.H.: Justice as a mediator of the relationship between methods of monitoring and organizational citizenship behavior. Acad. Manag. J. **36**, 527–556 (1993)
POWERGRID: Knowledge Management Policy. POWERGRID (2010)
POWERGRID: Knowledge Maps and Matrices. POWERGRID (2011a)
POWERGRID: Learner's Planner 2011-2012. POWERGRID (2011b)
POWERGRID: HRD Learner's Planner 2012-2013. POWERGRID (2012)
POWERGRID: HRD Learner's Planner 2013-2014. POWERGRID (2013)
POWERGRID: POWERGRID HRD Plan. POWERGRID (2014a)
POWERGRID: Annual Diary (2014b)
Steiger, J.H., Lind, J.C.: Statistically-based tests for the number of common factors. (1980)
Tucker, L.R., Lewis, C.: The reliability coefficient for maximum likelihood factor analysis. Psychometrika **38**, 1–10 (1973)
Van Dalen, D.B.: Understanding Educational Research: An Introduction. McGraw-Hill (1973)
Vigderhous, G.: The level of measurement and 'permissible' statistical analysis in social research. Pacific Sociol. Rev. **20**(1), 61–72 (1977)
Wheaton, B., Muthen, B., Alwin, D.F., Summers, G.F.: Assessing reliability and stability in panel models. In: Heise, D.R. (ed.) Sociological Methodology 1977, pp. 84–136. Jossey Bass, San Francisco (1977)
Wiersma, W.: Research Methods in Education: An Introduction. Allyn and Bacon (2000)
Winer, B.J., Brown, D.R., Michels, K.M.: Statistical Principles in Experimental Design, 3rd edn. McGraw–Hill, New York (1991)

Chapter 6
Discussion and Conclusion

6.1 Introduction

The companies have found it difficult to implement and maintain effective Knowledge Management programmes. A great interest in explaining this phenomenon has been the hallmark of the research on Knowledge Management. In order to find out how firms can realize the value that Knowledge Management promises while sidestepping the pitfalls, Gold et al. (2001) in their epoch-making article have argued that the problem of effective Knowledge Management stems from such organizations which do not consider their capabilities prior to undertaking a Knowledge Management Implementation programme. As a result of this, the guidance available for practitioners on implementing Knowledge Management is very little. The present study has provided an empirical validation that Knowledge Management can be implemented in a Company with many business verticals. The entire gamut of Knowledge Management from envisioning to evaluation has been provided in this study.

To test the identified hypothesized relationships of various variables and in order to answer the research questions, both qualitative and quantitative tools had been used. Research has suggested that an organization's culture is one of the most significant components of effective Knowledge Management and also one of the most difficult hurdles to overcome due to its complex nature. Gold et al. (2001) found that when culture is operationalized around the themes of corporate vision, corporate values and innovation, it is significant factor of an organization's infrastructure capability. In turn, Gold et al. (2001) found that infrastructure capability is a significant predictor of organizational effectiveness. At this juncture, it is not out of place to conclude our research in a sequential manner governed by the already defined research model. To evaluate these areas of interest of research, theories and prior research findings from several reference and discipline were researched and then combined through several iterations to formulate this intriguing view of the relationship between the factors of interest. In order to overcome this,

K.N. Pandey, *Paradigms of Knowledge Management*, Studies in Systems, Decision and Control 60, DOI 10.1007/978-81-322-2785-4_6

recourse was taken to random sampling so that the picture may emerge as a whole in totality to avoid any bias of age, defined level in POWERGRID, point of time of joining the Company and to cover all the five business verticals of POWERGRID.

6.2 Answering the Research Questions

The following four research questions as elaborated in Sect. 4.2 above had been identified for exploring them in detail through this research, and the findings have been delineated after every research question in subsequent paragraphs.

(1) ***How to examine whether knowledge is an entity as proffered by epistemological viewpoint or is it an organizational asset which can be managed?***

In order to answer this question, we collected data on three counts: (i) knowledge is treated as an asset; (ii) knowledge is treated as an object; (iii) knowledge is used as an important and critical resource. The mean rank of these three questions is 2.09, 1.84 and 2.07, respectively. It may be deduced from this that knowledge is an important asset which can be managed and only explicit knowledge can be treated as an entity as proffered by epistemologists. Even while interpreting the qualitative data, we found that knowledge is treated as an asset in POWERGRID. If Knowledge Management is to be implemented in letter and spirit in a Company and in the similar context, which has got various business verticals, it will have to treat knowledge as an asset and entity both. This finding is in line with the similar research carried out by Pangil and Chan (2014) where it has been concluded that knowledge is seen as a critical resource for competitive advantage.

(2) ***How to provide a holistic view of Knowledge Management Implementation to help theory-building effort of Knowledge Management Implementation to emerge?***

While analysing the qualitative data, we found that aspects such as Knowledge Management Cell, Knowledge Management Policy, Nodal Officers for Knowledge Management, Knowledge Management Awareness Programmes, Pilot Implementation of Knowledge Management Policy, Knowledge Bank, Task force for implementation of Knowledge Management, Knowledge Audit and Publication of Knowledge Maps and Matrices, Corporate Yellow Pages (Subject Matter Experts), Initiation of Communities of Practice, Training on Knowledge Management, Structure, Top Management Support, Reward, Incentive and Recognition, Mentor/Mentee Programme and Library and Information Centre are taken recourse to for the Knowledge Management Implementation and they have enabled POWERGRID to be a knowledge compliant Company.

The analysis of quantitative data of factor reductions has concluded that there are 15 factors: (1) IT-enabled KM activities; (2) recognition and contribution of best practices facilities; (3) intranet and Internet platforms; (4) ergonomic facilities for KM; (5) communications are facilitated; (6) top management support; (7) colloquium and seminar; (8) KM in actions; (9) unused knowledge assets; (10) KM: a way of life; (11) uses of knowledge tools; (12) periodical recognition for contribution; (13) learning from experience; (14) collection of knowledge and its capitalization; (15) KM-oriented mindset endorses that these quantitative constituents have helped the Knowledge Management in the Company (POWERGRID). Besides these 15 factors, these resemble the 14 assumptions that they are a reality.

Since both qualitative and quantitative data have concluded that Knowledge Management had been implemented in POWERGRID in letter and spirit both, it purports that the limitations of single case study may be overcome and the theoretical framework of POWERGRID may be used in similar context in other companies as well. In this context, it is found that the contour of schema depicts that the patterns and frameworks adopted in POWERGRID may also be adopted by similar organizations.

(3) ***How to identify the causal relation between affecting the Knowledge Management Processes and practices in POWERGRID with particular emphasis on technological parameters and HR parameters?***

On the basis of the reliability test, it can be concluded that the information and communication technology-related factors have got more consistency, and there are two reasons for this: (i) the Knowledge Management enables knowledge economy to flourish and (ii) there are globalization and virtualization taking place in an increasingly bigger and bigger way day by day. The sole reason is ubiquitous nature of information and communication technology which has pervaded the technical Company, i.e. POWERGRID more than other companies. Nevertheless, the human resource-related factors such as culture, training and development, reward and recognition, structure, team-based management and communication are equally important. The latest research on culture has depicted that 'most of us (especially knowledge workers) are energized by three fundamental things: autonomy—the freedom to direct our lives; Mastery—our craving to excel; and purpose—the yearning for our work to serve something larger than ourselves' (Fernández Aráoz 2014, p. 50). In this regard, the importance of top management support becomes pivotal, and besides finding the place in this study, it is also in line with the research reported by researchers (Hill et al. 2014; Christensen and Bever 2014); as part of this, the importance of communities of practice as found in POWERGRID is also pinpointed by researcher (Hill et al. 2014) when they underline the importance to build willingness by the top management.

While testing the hypotheses in predefined model, it was found that all the 12 hypotheses except one related with technology have emanated from human resource management and after testing all the 11 HR related hypotheses, all have been accepted (null hypotheses rejected). It brings to the fore that all HR-related Enablers are significant. It can be logically concluded that HR-related factors are more important being in majority than single factor related with information and communication technology. POWERGRID has rightly adopted all the 11 HR factors and single information and communication technology-related factor to fully implement the Knowledge Management across the length and breadth of the Company.

The findings of this research are in line with the findings of other researchers. Since the culture has superimposed all other factors in the predefined research model, we find that knowledge is a significant asset of the organization and it can be managed properly, provided the cultural aspects are taken care of from implementation point of view. Our finding pinpoints that culture is the prime mover of Knowledge Management Implementation and it is strongly related with the culture of the organization. This is in line with the findings of other scholars (Liebowitz and Chen 2001; Connell et al. 2014).

Some salient features of human resource management, such as training, structure and incentive, have also been endorsed by various researchers. The importance of training has been in line with the findings of other researchers like Jurgens and Krzywdzinski (2013), Zangiski et al. (2013) and Zarifian (2001). Organization structure as a facilitating construct has been reported by other researchers also in their finding (Gonzalez and Martins 2014). So far as the system of awards, rewards and incentives is concerned, the same have also been endorsed as HR factors helping Knowledge Management Implementation in a big way (Amayah 2013).

(4) ***How to test propositions of Knowledge Management Processes and practices with a view to provide a conceptual framework for Knowledge Management Implementation in a Company with many business verticals?***

This research was undertaken with a view to explore empirically the factors which are responsible for success of Knowledge Management in a Company with various business verticals. Apropos, this various hypotheses had been tested and summarized as given in Table 6.1.

The *p* value of all the null hypotheses is less than 0.05: it means null hypotheses have been **rejected** and all the above alternate hypotheses have been **accepted**. This test reflects that all the propositions have their significant role in Knowledge Management and create a strong conceptual framework for Knowledge Management Implementation in the Company with defined five business verticals. It will be applicable for all other companies/industries those who have many business verticals with multilayer implementation of Knowledge Management.

These research questions had been delved and pondered over while the research was in progress. The predefined business model helped in answering the research questions. The same model had been applied by POWERGRID while undertaking the initiative of knowledge reproduced.

6.3 Status of Knowledge Management in POWERGRID

All along the growth path of business of POWERGRID, information, as a vital input, has been playing crucial role along with man, material, money and method in the process of establishment and maintenance of a vast pool of transmission infrastructure spread over multiple locations across the country. While old IT infrastructure has been progressively updated to keep pace with technology

Table 6.1 Results of Hypotheses

	Hypothesis (H_a, alternative hypothesis)	Tests	Results
H_1	Knowledge as an asset can be managed	Related sample test (Friedman test)	Accepted
H_2	The attitude and behaviour of personnel help Knowledge Management Implementation positively	Related sample test (Friedman test)	Accepted
H_3	Culture is the prime mover of Knowledge Management Implementation	Related sample test (Friedman test)	Accepted
H_4	Strategy directs Knowledge Management Implementation favourably	Related sample test (Friedman test)	Accepted
H_5	Team-based matrix structure affects Knowledge Management Implementation	Related sample test (Friedman test)	Accepted
H_6	Well-laid systems enable Knowledge Management Implementation suitably	Related sample test (Friedman test)	Accepted
H_7	Predefined processes help Knowledge Management Implementation	Related sample test (Friedman test)	Accepted
H_8	Management support is a prerequisite for Knowledge Management	Related sample test (Friedman test)	Accepted
H_9	Relevant training to personnel is required for Knowledge Management Implementation	Related sample test (Friedman test)	Accepted
H_{10}	Reward and recognition are motivators for effective Knowledge Management	Related sample test (Friedman test)	Accepted
H_{11}	Information and communication technology is an enabler for Knowledge Management Implementation	Related sample test (Friedman test)	Accepted
H_{12}	Tools related with HR management are helpful for effective Knowledge Management	Related sample test (Friedman test)	Accepted

advancement, information system continues to work primarily in a disintegrated structure. Today, all departments have turned into a large storehouse of functional information, and the organization as a whole is now a vast repository of knowledge primarily embedded within individual brains. However, need for information decision is being felt more and more at organizational level which is proving difficult if elements of information are available in isolation. Multiple platforms for applications and databases for information system exist for diverse functions which pose problems for cross-functional information exchange, hence the need for Enterprise Resource Planning. Business managers often need information that cuts across functional silos and quality of deploying a Knowledge Management System based on Enterprise Resource Planning on a single platform that will act as a single source to access, manipulate, create and update information for use of all knowledge workers.

Knowledge maps and matrices have proved to be an important document which personnel of POWERGRID are using greatly.

Corporate Yellow Pages has become a document of reference, and employees are referring to it to establish contact with subject matter experts for availing tacit knowledge.

Affinity Members and Domain Leaders are yet to be identified in POWERGRID. It is largely untouched because there is no chief knowledge officer to spearhead this task as of now in Company. The contributors to knowledge topics may be a group of employees or an individual employee interested in a particular topic. Such contributor(s) will be called 'Affinity Members'. Domain Leaders, who will be volunteers, will add values to the contributed pieces of information and knowledge in the knowledge bank of his community of practice. The Domain Leaders will not be limited to a particular location but could be located in any of the POWERGRID's establishments. Domain Leaders shall generally be persons who have an in-depth knowledge about the topic. However, an employee who collects knowledge from other employees and contributes shall also be recognized as Domain Leader.

Virtual KM organization is yet to take roots. Nurturing KM habits, particularly identification, acquisition, sharing and preserving knowledge, should be everyone's responsibility in POWERGRID. The Knowledge Management Cell comprises of professionals from electrical engineering, electronic engineering, human resource and information technology. The cell is responsible for forming procedures and managing the KM process in coordination with the various KM Nodal Officers. The cell also facilitates and monitors the Knowledge Management usages/practices in the organization. Thus, KM organization has become a virtual organization.

Incentive and recognition are in poor shape as the scheme has not been fully implemented.

Heads of departments, regions and RLDCs are to be sensitized for inculcating Knowledge Management culture, especially sharing. They are also to glean as which knowledge should go in public domain and which knowledge should not be made public.

Knowledge Management Cell should be concentrated into a full-fledged department by strengthening it with requisite staff. This will help ushering in knowledge culture in POWERGRID which is in its infancy.

Knowledge Management is in full swing in the grid management business of POWERGRID. Telecom business vertical is very close to grid management so far as Knowledge Management is concerned. Operation and maintenance of transmission assets business is also doing very well. These three business verticals have been using existing information technology. However, project management and consultancy business of POWERGRID are also trying to come up at the level of other three businesses.

It was noted that POWERGRID was using its existing knowledge in new businesses in a big way which is a real capitalization of knowledge; for example, in Ethiopia Consultancy Project, they prepared a document entitled 'Management Knowledge, Improvement and Change' which included developing enterprise-wide Knowledge Management capability, developing KM strategy, assessing KM approaches, planning and designing of KM projects, managing KM projects, measuring KM projects and monitoring KM performance. In fact, the Ethiopian

Consultancy Project was installed as a Knowledge Management project. This is the newest project of POWERGRID.

One thing is of special mention here that POWERGRID has implemented its Knowledge Management which is 'home-grown' and they did not hire any consultant for that from outside.

POWERGRID undertook the capitalization of knowledge through its business vertical, namely *Consultancy*.

POWERGRID has taken lead role in implementing *National Knowledge Network* project of Government of India.

6.4 Supremacy of HR Apropos Technology

The research has established that the model which was adopted by POWERGRID had only one component of information and communication technology and the rest of the parameters were from the domain of human resource management including the superimposing position of culture. It purports that any firm which intends to take cues from the model adopted by POWERGRID should entrust the initiation of Knowledge Management to human resource department. Further, the problem is that the people who have important information do not possess the desire to convey the information and those who need the information and knowledge do not get the means to receive and use the knowledge. 'Fortunately, HR can play an active role in addressing the issues of technology and information management, adding materially to its value to the business' (Ulrich et al. 2012, p. 184). Nevertheless, the role of information and communication technology as an enabler is underlined especially in an era of dominance of electronics-enabled media.

6.5 Implications for Management and Researchers

This research has got implications for managers and researchers alike. The factors comprising of (1) IT-enabled KM activities; (2) recognition and contribution of best practices; (3) intranet and Internet platforms; (4) ergonomic facilities for KM; (5) communications are facilitated; (6) top management support; (7) colloquium and seminar; (8) KM in action; (9) unused knowledge assets; (10) KM: a way of life; (11) usage of knowledge tools; (12) periodical recognition for contribution; (13) learning from experience; (14) collection of knowledge and its capitalization; and (15) KM-oriented mindset can be used as tools by managers for the implementation of Knowledge Management in their companies and organizations. The researchers can take support from these factors as it is empirically explored and conclusively proved. Information technology (IT), supports and facilitates the rapid collection; storage, and exchange of knowledge on a scale not practicable previously, thereby assisting in the knowledge creation process, a well-developed system or technology which helps to integrate fragmented flows of knowledge, and can eliminate barriers to communication and facilitates all modes of knowledge creation. While explicit knowledge is comparatively easy to store and communicate in any form, it does present fewer problems being stored, centrally, and transmitted via information systems.

In this study, various variables such as culture are ascertained: knowledge as an asset can be managed; the attitude and behaviour of personnel help Knowledge Management Implementation positively; culture is the prime mover of Knowledge Management Implementation; strategy impacts Knowledge Management Implementation directly; team-based matrix structure affects Knowledge Management Implementation; well-laid systems enable Knowledge Management Implementation suitably; predefined processes help Knowledge Management Implementation; management support is a prerequisite for Knowledge Management; relevant training to personnel is required for Knowledge Management Implementation; reward and recognition are motivators for effective Knowledge Management; information and communication technology is an enabler for Knowledge Management Implementation; tools related with HR management are helpful for effective Knowledge Management which helped Knowledge Management Implementation in POWERGRID.

6.6 Limitations of This Research

This research focused on the Knowledge Management in its entirety in POWERGRID by hypothesizing various variables. There are, doubtless, other variables that influence the Knowledge Management activities in a Company. It is entirely possible that other or additional environmental factors may be determined to possess a more direct or powerful influence on the schema presented in this research.

The second limitation of this study is the assumption that only executives are knowledge workers in POWERGRID and other two categories of supervisors and workmen are not 'knowledge worker' and hence have been excluded.

The third limitation of the research is its bias towards sampling of English-knowing personnel. The fourth limitation was posed by time constraint which could not allow reaching out to the entire 'population'.

A broader sample of employees covering all the three categories including supervisors and workmen besides executives would have removed any potential bias that may exists due to the participation of only executives in the survey.

The cultural differences could also possibly exist in this study because the data were collected across the length and breadth of POWERGRID which has got its presence across entire India and beyond. The findings of this study should be interpreted cautiously because a single organization has been studied, and the framework and the schema of this study should be seen as organization specific.

Although content analysis is a useful tool for analysing qualitative data, issues arose to its use in this research project because the personnel involved in Focus Group and interview could be taking POWERGRID in high esteem and there is a possibility of existing of gap between what is said and what exists on the ground.

6.7 Future Research

Based on this study and its findings, there are several areas for future research: first and foremost, the study of Knowledge Management should be carried out on the level of inter-organizational and firms in power industry so that generalizability could be possible and a model could be thrown up. Besides this, future research could be undertaken to establish a better understanding of the factors which enable or do not enable a Knowledge Management Implementation. This research is based on the views expressed by the executives and their experiences apropos POWERGRID—it would be valuable to gather views and experiences of a wider range of people both internal and external to POWERGRID. Given the case study method of this research, it is recommended that broader, more internationally generalizable research is conducted that includes cluster firms within a range of sectors.

6.8 Conclusion

The Company selected for this research through a single case study is a Company with 5 business verticals and is having all India presence besides its footprints outside India also. Considering the aim of this study, an explorative, qualitative and quantitative research approach appeared to be more appropriate. The outcome of research has brought 'Knowledge Management' in centre through the mixed methodology used in this research covering the entire odyssey of Knowledge Management in POWERGRID from envisioning to evaluation and capitalization. The framework and schema of Knowledge Management presented in this study could be generalized as an outcome of case study (in accordance with Viale (2010)). The case study has been an important tool for empirical enquiry. It has rightly been pointed out by Yin (2003) that 'A case study is an empirical enquiry that investigates a contemporary phenomenon in-depth within its real-life context, especially when the boundaries between the phenomenon and context are not clearly evident' (p. 18). According to the definition from Yin (2003), the case study method focuses on the meaning of the research design. It is believed that this study has answered all the research questions and has contributed to the literature of Knowledge Management, especially in its entirety from envisioning to evaluation and capitalization.

Nine factors from human resource management and two factors form information technology, totally to eleven, have helped POWERGRID in implementing Knowledge Management. It implies that it will lead better strategic Knowledge Management for the Company. It may be applicable to other organizations which have similar business verticals.

References

Amayah, A.T.: Determinants of knowledge sharing in a public sector organization. J. Knowl. Manag. **14m**(1), 100–113 (2013)

Christensen, C.M., Bever, D.V.: SPPTLIGHT: Are investors bad for business. Harvard Bus. Rev. 53–70 (2014)

Connell, J., Kriz, A., Thorpe, M.: Industry clusters: an antidote for knowledge sharing and collaborative innovation? J. Knowl. Manag. **18**(1), 137–151 (2014)

Fernández Aráoz, C.: 21st Century talent spotting. Harvard Bus. Rev. 42–52 (2014)

Gold, A.H., Malhotra, A., Segars, A.H.: Knowledge management: an organizational capabilities perspective. J. Manage. Inf. Syst. **18**(1), 185–214 (2001)

Gonzalez, R.V.D., Martins, M.F.: Mapping the organizational factors that support knowledge management in the Brazilian automotive industry. J. Knowl. Manag. **18**(1), 152–176 (2014)

Hill, L.A., Brandeau, G., Truelove, E., Lineback, K.: Collective genius. Harvard Bus. Rev 88–97 (2014)

Jurgens, U., Krzywdzinski, M.: Breaking off from local bounds: human resource management practices of national players in the BRICS countries. Int. J. Automot. Technol. Manage. **13**, 114–133 (2013)

Liebowitz, J., Chen, Y.: Developing knowledge sharing proficiencies. Knowl. Manag. Rev. **3**(6m), 12–15 (2001)

Pangil, F., Chan, J.M.: The mediating effect of knowledge sharing on the relationship between trust and virtual team effectiveness. J. Knowl. Manag. **18**(1), 92–106 (2014)

Ulrich, D., et al.: HR from the Outside In: Six Competencies for the Future Human Resource. Tata McGraw Hill, New Delhi (2012)

Viale, R.: Knowledge driven capitalization of knowledge. In: Viale, R., Etzkowitz, H. (eds.) The Capitalization of Knowledge. Edward Elgar, Cheltenham, U.K. Northampton, M.A., USA (2010)

Yin, R.K.: Case Study Research: Design and Methods, 3rd edn. Sage Publication Inc., Thousand Oaks, California (2003)

Zangiski, M.A.S.g., Lima, E.P., Coasta, S.E.G.: Organizational competence building and development: contributions to operations management. Int. J. Prod. Econ. 144, 76–89 (2013)

Zarifian, P.: Objectivo Competencia: por uma nova logica. Editora Atlas, Sau Pualo (2001)

Appendix A
Focus Group Discussions with Nodal Officers of Knowledge Management of POWERGRID on 27.01.2014/28.01.2014

A.1 Moderator—Shri Yatindra Dwivedi, Chief Manager (HR) POWERGRID

The focus group discussion took place for 1½ h each on 27.01.2014 and 28.01.2014 respectively inviting Knowledge Management Nodal Officers of POWERGRID from various departments, all five business verticals, Regional Headquarters and Regional Load Dispatch Centres. The focus group discussions were conducted with a view to elicit in-depth information of knowledge management from envisioning to evaluation. The attempt was made to trace the role of Nodal Officers in knowledge management in POWERGRID and their activities. Following points emerged from the discussions:

- POWERGRID is an organization which is structured on bureaucratic lines. Therefore, it is hierarchical in its structure. Such an organization requires guidelines from top. By keeping this in view, Knowledge Management Policy was formulated and approved by Board of Directors.
- Once the Knowledge Management Policy was in place, it was piloted in Southern Region-II and once the pilot implementation of policy was done, the policy was revised in 2010 to remove the anomalies noticed during pilot. The revised policy was afterwards implemented throughout the length and breadth of company.
- The policy envisaged identification of Nodal Officers in every department of corporate centre, every region of POWERGRID and every Regional Load Despatch Centre. Care was taken to cover all the five business verticals viz. Operation and Maintenance of transmission assets, Project Management, Telecom, Grid Management and Consultancy.
- The role and responsibility of Nodal Officers include:
  - Responsibility for preparing the knowledge map that defines knowledge needs, knowledge assets and its categorization, knowledge owners and explicit contents, and knowledge users, and updation of this map on regular basis.

K.N. Pandey, *Paradigms of Knowledge Management*, Studies in Systems, Decision and Control 60, DOI 10.1007/978-81-322-2785-4

  - Accountability for improving the circulation of knowledge throughout the organization.
  - Responsibility for creating awareness and conducting at least two knowledge sharing sessions per year.
  - To identify Knowledge Agents, if required to facilitate knowledge management at spread out locations.

- Nodal Officers prepared Knowledge Maps and Matrices of POWERGRID and published it in January, 2012. This document helped company in formulating the strategy, knowledge requirements, knowledge gap, identifying existing knowledge assets of various domains.
- Knowledge Maps and Matrices clearly delineated what knowledge is needed, who has this knowledge, who needs this knowledge, where is this knowledge available, is it tacit or explicit, is it routine or non-routine and which issues does it address.
- It was found by Nodal Officers that employees wanted to know the subject matter experts so that they could directly contact them for their knowledge requirements. As a result of this Corporate Yellow Pages were prepared through inviting the employees to declare their expertise in a particular field voluntarily. Once these declarations were available Knowledge Management Cell of POWERGRID collected it and electronically published it in 2013 so that these could facilitate free-flow of knowledge between 'knowers' and those who needed it.
- A workshop was conducted to initiate the Communities of Practice (CoPs) in 2012 and anchors were identified for communities of Operations and Maintenance of transmission assets, Telecom, Project Management, Consultancy and Grid Management. CoPs were also formed for enablers like Human Resources, Finance and Information Technology.
- POWERGRID is undertaking knowledge management activities like a project since projects are the formal mechanism for change in the company. They follow a formal process for project management that addresses initiation, execution and completion of a part of the whole journey of knowledge management in POWERGRID.
- They have created a successful knowledge management environment in POWERGRID. Their next step is to establish enterprise knowledge goals and group knowledge objectives that articulate what the total organization expects to gain from knowledge management.

Appendix B

B.1 In Depth Interview of Knowledge Management Implementation Committee of POWERGRID on 17.01.2014

Researcher	What is the requirement of Knowledge Management Implementation Committee in POWERGRID?
Shri R.K. Tyagi	The purpose of the formation of Knowledge Management Implementation Committee is to fulfill the objectives of the Knowledge Management Policy of the Company
Researcher	Will you please elaborate these objectives?
Ms. Pratyksha	Formulation of Knowledge Management Policy had been done in POWERGRID to ensure: • Harnessing potential of Information and Communication Technology • Increasing the organizational process efficiency and accuracy for maximizing the returns • Enriching knowledge content for establishing and creating new ways and means in business process of organization • Facilitating and management of organizational innovation and learning • Leveraging the expertise of people across the organization • Increasing the knowledge exchange among employees • Managing intellectual capital and intellectual assets in the work force • Adopting new knowledge management tools, techniques and technologies to keep the KM system in POWERGRID intact, robust and going • Increasing employees satisfaction by facilitating employee learning
Researcher	Which process did you undertake to implement knowledge management in POWERGRID?
Mrs. Kumud Wadhwa	The broad processes adopted for managing knowledge in POWERGRID had been conceptualized into following four stages: • Create/Capture—Identify the required knowledge domains in alignment with KM strategy, locate the source of this knowledge (internal/external), and acquire or generate the required knowledge • Collate—Classify/codify the knowledge objects (documents and create knowledge-base), synthesize (seek patterns across different knowledge

(continued)

K.N. Pandey, *Paradigms of Knowledge Management*, Studies in Systems, Decision and Control 60, DOI 10.1007/978-81-322-2785-4

(continued)

	objects), identify the target group for the different knowledge objects, represent (refine, organize and present the knowledge objects in a user-friendly manner), and adapt (translate the knowledge objects to the local context) • Share—Implement and maintain knowledge sharing systems on KM Portal (like groupware, bulletin boards, meeting, etc.) and disseminate (to the relevant target groups) • Capitalize—Monitor usage, assess/measure the benefits and the knowledge management in terms of the specified knowledge goals, get feedback, review and renew the knowledge bases (including identification of new knowledge to be captured), and embed knowledge into the organization's value creating activities (services and/or information)
Researcher	What structure did you follow for knowledge management in your company?
Shri Yatindra Dwivedi	The structure we followed was to serve on two counts—explicit and tacit • Explicit knowledge has been captured by designated persons of respective functions. To elaborate, if a circular from Contracts Services (CS) department is an identified knowledge assets as per knowledge map, the concerned executive from contract has been designated and authorized to upload those circulars on knowledge portal • Tacit knowledge may be captured by any employee in our company. The employee will post his/her report, findings collections or write-up to the appropriate topic/classification directly in the knowledge bank. The sharing of the tacit knowledge will be basically informal sharing of experiences and shall not be the only basis of taking a decision. For example, it was found that Low frequency heating of transformer coil could not produce positive results in one case, however based on further study and findings of international research low frequency heating was still done in another transformer and it produced positive results. Tacit Knowledge Sharing shall consider the contextual factors. This new contribution shall be open for review/comment by any of the employees of POWERGRID. Any suggested modification will be appended to the topic. The Domain leaders will then appropriately classify the topic, modify it and include in the formal Knowledge Bank. The Affinity Member will be the owner of the information knowledge/Topic contributed by him
Researcher	What steps did you take to fructify these processes?
Shri Arun Kumar Singh	Knowledge Bank has been established where following contents find place and it is kept updated from time to time: • People Search—Expert pages (A list of domain leaders), Project owners, Vendor information, Customer information • Training material—FAQs, Presentation, Audio/Video files, Orientation courses, reading material, Self-study and evaluation courses for a specific role. • Literature—Books/Manuals, White papers/Research articles. • Discussion forums/Blogs • Lessons Learnt—known issues and workarounds for Project Issues as well as Maintenance Issues, experiential knowledge

(continued)

(continued)

	• External Resources—Link of external websites for reference, Book Reviews, External Environment Inputs like Regulations etc., Innovations/Latest Technology • Corporate Information—Progress Reports, MIS Reports as given in department ISO procedures, measures and matrices • Project information—Progress Reports, MIS Reports as given in departmental ISO procedures; Measures and Matrices • Technical Drawing, specification and blue prints for various components, equipments and systems • Technical Reports—Test Results, Outage Reports, Failure Analysis Reports, Assessment reports • Records of meeting—PRM/ORM Agenda and Minutes, Vendor's meet Agenda/Minutes, Minutes of PNBC Meeting, Minutes of HOP meetings • Competitor's Information • Bazzwords • Glossary We adopted the mechanism for collation, updation, use and reuse of the contents of knowledge bank. Once we published knowledge maps and matrices and corporate yellow pages we could take care of knowledge on the level of explicit and tacit both
Researcher	How did you ensure the availability of tacit knowledge of subject matter experts to those who need it?
Dr. Sunita Chauhan	We prepared corporate yellow pages which contains the list of subject matter experts in the areas of Operation and Maintenance of Transmission Assets, Telecom, Grid Management, Project Management, Consultancy, Vigilance, Common-Collaboratory of Core Expertise, HR, Finance, IT and Miscellaneous. We delineated the professional area, area of specialization, types of expertise, while giving the employee no., name, designation, e-mail, mobile number of experts. In addition to that we also gave a brief description of experience/exposure, problems faced/solutions offered/tools and drivers used by the subject matter experts in corporate yellow pages. This facilitated our employees to contact subject matter expert directly without having any activity related with gate keeping
Researcher	Have you got the support of top management for implementation of knowledge management in real sense of the term?
Ms. Minaxi Garg	Yes, we have got whole hearted support of our CEO (Chairman and Managing Director) and other Board Members. The role of CEO in culture change for knowledge management has been immense. In fact, employees in POWERGRID find the top management supporting and promoting knowledge sharing from top to bottom
Researcher	Do you find your human resources department functioning favourably for implementation of knowledge management?
Shri Arun Kumar Singh	Human Resource Department took the initiative to implement knowledge management in POWERGRID. They got the Knowledge Management Policy formulated and subsequent actions started
Shri Yatindra Dwivedi	They went well beyond official requirement to conduct Knowledge Management Awareness Programs across POWERGRID which created ripple effects and helped formulating a knowledge sharing culture

(continued)

(continued)

Shri V.K. Singh	Not only this, Human Resource Development Department has been enabling the learning, especially for knowledge management, through it's offering of Knowledge Management Programme in its annual planner. This has been taking place for last three years consecutively and those desirous of improving their knowledge have got themselves enrolled as participants
Researcher	Do you find the existing Information and Communication Technology adequate for implementation of knowledge management in POWERGRID?
Mrs. Kumud Wadhwa	Initially, the existing Information and Communication Technology was not adequate. However, we are setting the requisite technology through installation of Enterprise Resource Planning which has enabled knowledge acquisition, generation, dissemination possible
Researcher	How do you measure the process of implementation of knowledge management in POWERGRID?
Shri Anand Shankar	There are three factors through which we measure the progress. One is the number of hits in our Knowledge Bank. Second is the number of queries which are made to the subject matter experts and third is the number of participants who attend the knowledge management programs and workshops
Dr. Sunita Chauhan	Even the number of employees who visit our library is an indicator of the progress of knowledge management implementation in POWERGRID
Shri Yatindra Dwivedi	Sometimes, the request for associating of specialists in Cross Functional and Internal Departmental Teams also provides us the measuring rod of knowledge management implementation in company
Researcher	Do you also associate outside agencies in your knowledge management implementation?
Mrs. Kumud Vadhwa	We send our personnel to other companies such as NTPC, SAIL, BHEL and Coal India to study their best practices of knowledge management and we implement them in POWERGRID
Researcher	Will you please indicate what actions you are planning to take for betterment of your knowledge management implementation in times to come?
Shri Prashant Kunaujia	On the front of explicit and tacit knowledge we have a sense of satisfaction. So far as the implicit knowledge of our retiring employees is concerned, we are developing a plan to capture their knowledge as well before they depart from POWERGRID

Appendix C

C.1 Elite Interview of Director (Personnel) Shri Ravi P. Singh of POWERGRID on 16.01.2014

Researcher	Sir, Thank you very much for sparing your precious time for this interview
Shri Ravi P. Singh	It is O.K., I believe this will help me also in fine tuning my vision and strategy for knowledge management in my company
Researcher	Can you give an outline of the knowledge management as you have visualised it for POWERGRID?
Shri Ravi P. Singh	I am aware that the globalization of my company has brought in knowledge economy in which the challenge is to build, combine and integrate the knowledge assets of the more than ten thousand employees so that this asset can be used, reused and capitalized gainfully
Researcher	What strategy do you adopt for knowledge management?
Shri Ravi P. Singh	Knowledge management strategy is part of the overall business strategy of POWERGRID which is the responsibility for planning and coordination of interstate transmission system since we have one grid for entire India. Besides this, we are the major player in establishing SAARC grid. On the global front, we have our businesses in Bangladesh, Bhutan, Dubai and Ethiopia and businesses in other countries are in pipeline. So far as knowledge management is concerned, POWERGRID has been given the responsibility of 'implementing agency' for National Knowledge Network on high speed connectivity through my telecom network. Therefore, my knowledge management strategy enables dissemination and circulation of knowledge among POWERGRIDIANS to gain an organizational competence and to incorporate this knowledge into organizational processes
Researcher	How do you translate your strategy in practice?
Shri Ravi P. Singh	I have created dedicated teams to acquire, create, share and use knowledge strategically in POWERGRID for enabling it to become a pioneer company in all its business domains
Researcher	Do you provide adequate resources to your teams?
Shri Ravi P. Singh	Yes, more than adequate resources in terms of money, time and manpower have been provided to the teams entrusted with the knowledge management work

(continued)

K.N. Pandey, *Paradigms of Knowledge Management*, Studies in Systems, Decision and Control 60, DOI 10.1007/978-81-322-2785-4

(continued)

Researcher	I am sure, you lead your personnel with vision, grit and determination, would you please like to elaborate you themes of knowledge sharing culture?
Shri Ravi P. Singh	For an organization to adopt knowledge management successfully, the practice of knowledge sharing must be integrated into the day-to-day routines of organization's employees. In POWERGRID we have a work culture where collaboration is pivotal among employees and knowledge sharing is not an exception. In fact, we incentivize knowledge sharing through recognition which is the prime mover of the knowledge sharing culture
Researcher	You are part of the Board of Directors, in this respect how you ensure continuance of knowledge sharing culture?
Shri Ravi P. Singh	We provide job security to people so that they feel their job is secured that is why they do not get scared in sharing knowledge. Thus they do not feel that once knowledge is shared they may become redundant and sparable. POWERGRID has placed conducive structure and free flow of knowledge is an outcome of job security and proper structure of work flow
Researcher	Do you derive competitive advantage from your knowledge assets?
Shri Ravi P. Singh	We have knowledge asset in abundance in all our business verticals. In acquiring these rare assets we could make them rare because of our hard work. POWERGRID has succeeded in identifying its efficiency and effectiveness and our ability in reusing them provides us competitive advantage, more so, through our human capital
Researcher	What is your opinion about learning especially in the context of knowledge management?
Shri Ravi P. Singh	Knowledge management is not only about managing knowledge but also about changing culture to one that values learning and sharing, and it facilitates the learning process. I ensure that the learning of employees has been given adequate component of knowledge management continuously. The flow of directives made up for knowledge, cognitive models and visions are expressed by the managerial team and the same is crucial to support the evaluation of individual and collective knowledge. The organizational learning includes the temporal requirement of refreshing the learning for updation of knowledge
Researcher	What is your future vision of Knowledge Management in POWERGRID
Shri Ravi P. Singh	My vision of knowledge management is POWERGRID will be knowledge provider to further the capabilities of its personnel for creating knowledge sharing culture to increase business in all its verticals by reducing time and cost
Researcher	What is your planning for future of knowledge management for your company?
Shri Ravi P. Singh	I am planning to obtain ISO 25006 certification for making POWERGRID as knowledge management compliant company
Researcher	Sir, thank you very much indeed

Appendix D
Questionnaire

D.1 Survey of Knowledge Management in POWERGRID

Dear Sir/Madam,

Greetings to you. As a working professional and expert in your field of activity in POWERGRID, you would be aware of the importance attached to the management of knowledge in Company. I am pursuing my research study, "Knowledge Management: A Case Study of POWERGRID". Knowledge Management (KM) is an umbrella term which refers to any deliberate effort to manage the knowledge of an organization's workforce, which can be achieved via a wide range of methods including directly, through the use of particular types of Information and Communication Technology, or more directly through the management of social processes, the structuring of organizations in particular ways or via the use of particular culture and people management practices. KM deals as much with people and how they acquire, exchange and disseminate knowledge as with information technology. As part of this research, I need to survey to trace the practice of knowledge management in POWERGRID from envisioning up to the evaluation of knowledge management in our Company.

A questionnaire dealing with various aspects of knowledge management is enclosed. Most of the questions pertain to the work areas of POWERGRID and its five business verticals—operation and maintenance of transmission assets, construction of projects and their management, grid management, telecom and consultancy covering the knowledge and its role in them and their enablers like HR, Finance, I.T., ESMD, etc. You are ideally suitable to respond to this questionnaire from your perspective of knowledge management in areas under your control and influence or the one in which you are working. This survey will give a fair view of knowledge management in POWERGRID.

I assure you that your response will be used for the purpose of research and will remain confidential. Thank you very much for your participation.

(K.N. Pandey)
Chief Manager (HRD)
POWERGRID, Corporate Centre
Gurgaon

K.N. Pandey, *Paradigms of Knowledge Management*, Studies in Systems, Decision and Control 60, DOI 10.1007/978-81-322-2785-4

D.2 Personal Informations

(Please put (√) mark in the box of your choice)

1. Name (optional) :

2. Gender: Male □ Female □

3. Age:
   a) 23-30 years □
   b) 31-38 years □
   c) 39-46 years □
   d) 47-54 years □
   e) 55 years & above □

4. Please mention your highest educational qualification :
   a) Graduation □
   b) Post-Graduation □
   c) Doctorate □

5. Your current level in POWERGRID :
   a) E1-E3 □
   b) E4-E5 □
   c) E6-E7A □
   d) E8 & above □

6. Department _______________

7. Total industry/work experience :
   a) 0-7 years □
   b) 8-14 years □
   c) 15-21 years □
   d) 22-28 years □
   e) 29 years & above □

8. Have you worked in any other public sector organization previously ?

   Yes □ No □

9. Year of joining POWERGRID ______

10. Are you associated with Knowledge Management work

    Yes □ No □

D.3 Questionnaire

Scale: **SD** (Strongly Disagree), **D** (Disagree), **NA/DA** (Neither Agree nor Disagree) **A** (Agree), **SA** (Strongly Agree)
Please put a (√) mark in the box of your choice.

		SD	D	NA/ND	A	SA
1	Knowledge Management has taken place in POWERGRID					
2	In my department, knowledge management activities are taking place satisfactorily					
3	In my opinion, there are very few Subject Matter Experts available in my department					
4	I have come across situations in work where the same mistake done by someone in the past is being repeated					
5	There are lots of documents/reports lying unutilized in my department					
6	Most of the time I find that it is pretty time consuming to access any document in my department					
7	There are appropriate values and mind set towards knowledge management amongst employees of my Company					
8	The personnel in POWERGRID are oriented towards knowledge management very positively					
9	Aesthetically oriented culture for knowledge management is a way of life in POWERGRID					
10	Employees in POWERGRID are having informal, open-minded and collaborative working					
11	Knowledge is treated as a resource/asset in POWERGRID					
12	Knowledge is treated as an object in my department					
13	Knowledge is used as an important and critical resource in my department					
14	POWERGRID has got a vision in knowledge management which properly articulates its knowledge needs					
15	Knowledge Management is being used constantly as a business strategy					
16	Transfer of Knowledge and Best Practices are taking place in day-to-day working of employees					
17	Innovation and Knowledge Creation are a way of life in my Company					
18	Reward and recognition for suggestion of Best Practices, have been incorporated in POWERGRID's annual plan					
19	Information about the outcome of implementing Best Practices is translated in practice in my department/company					
20	Dedicated staff for Knowledge capture and acquisition are earmarked in my department					
21	Information regarding relevant topics are readily available to the employees in my Company					

(continued)

(continued)

		SD	D	NA/ND	A	SA
22	Information regarding process description can be uploaded in the Organization's Knowledge Bank at Power Portal, the intranet of Company					
23	'Personal best practices can be shared with other employees easily' can be said as a part of job in POWERGRID					
24	A cohesive and powerful vision of the organization is shared across the workforce to promote need for strategic thinking at all levels					
25	The organization has enabling structures in terms of hierarchy and communication flows that facilitates knowledge sharing					
26	The organization has cohesive teams which facilitate sharing of experience and information amongst employees					
27	The organization provides incentives to motivate users to learn from experiences and use the Knowledge Management System of POWERGRID					
28	The organization continuously strives for Knowledge acquisition, dissemination and capture for its employees					
29	Employees are provided 'instant messenger' or 'chat'—video conferencing facilities through which they can have conversations with their colleagues					
30	The Intranet site provides an Online Discussion Forum in which employees can share information and post their doubts					
31	My organization has an Experts Database that can be used by Employees to locate experts within the organization					
32	Employees are provided the 'blogging' platform wherein they can share their experiences through blogs					
33	Employees can upload their documents or reports to Power Portal					
34	Employees can download any document they want from the Intranet site/portal					
35	My organization has a 'Chief Knowledge Officer' (or equivalent) who is responsible for managing the knowledge-related activities					
36	I visit Power Portal for my knowledge needs					
37	I use Knowledge maps and matrices of my Company for knowledge requirements					
38	I contact subject matter experts by using my Corporate Yellow Pages					
39	Our Company has got collegial, flat, team based, open door policy					
40	We strive to have matrix form of structure of Company					
41	Brick-and-mortar model (dealing with the customer face-to-face) structure is followed in our Company					
42	Every employee is imparted training required by him for his job					
43	Knowledge Management related training is identified as part of organizational training need assessment					

(continued)

(continued)

		SD	D	NA/ND	A	SA
44	My organization has a central repository/library where all the necessary documents/reports are stored					
45	All important documents in my organization are available in the electronic format					
46	Documents pertaining to past research programs are arranged in such a manner that they can be easily accessed and used as and when necessary					
47	There are enough Meeting/Conference rooms available for Team meetings and discussions					
48	Employees/Teams can discuss work through Tele-Conferencing (i.e. through telephones, VOIP, Mobile, etc.)					
49	All programs have periodic team meetings to review the progress and discuss issues pertaining to the work					
50	In my department/organization, every program has a closure meeting to critically review the performance					
51	Within the team/department, employees share their knowledge in an informal manner					
52	There is very good interaction between departments/groups in my organization					
53	My organization's newsletter/bulletin will inform me about the latest happenings in your organization and outside					
54	My organization's intranet/website plays a vital role in disseminating information to the employees					
55	All important communications in my organization are sent to me through circulars or memos in electronic environment					
56	Notice boards kept at various vantage points in my organization plays an important role in informing me of various news and events					
57	My organization has a directory of personnel containing information of names of people, their designation, division, program, phone number, email, etc.					
58	Information about locations of project reports, research facilities and equipments available are displayed prominently in my organization					
59	I contact Communities of Practice in my area and I am a member of these					
60	Knowledge is created in Technology Development Department and in Sub-stations of POWERGRID					
61	Acquisition of knowledge is a regular practice in our Company					
62	Use of knowledge is a way of work life in every employees' job					
63	Knowledge Management Cell of POWERGRID collates knowledge nuggets and disseminates it through Power Portal					
64	The capitalization of knowledge of Company is realized through our Consultancy Department					

(continued)

(continued)

		SD	D	NA/ND	A	SA
65	In my organization, every employee is encouraged to periodically produce *documents* in the form of technical reports, journal papers, presentations, articles, etc.					
66	In my organization, it is mandatory for every employee to document insights (*in the form of best practices, lessons learnt etc.*) from the work/project					
67	My organization sponsors trips for employees to international/national conferences and seminars					
68	My organization arranges seminars/lectures by external experts on a periodic basis					
69	All employees in my department are willing to share their knowledge with others					
70	I have free access to any information I want irrespective of where it resides in my organization					
71	I feel that I have got support from Management for knowledge management					
72	My superiors fully endorse my way of handling knowledge management related issues					
73	The Top Management supports and encourages Knowledge Management					
74	My organization periodically gives special awards to employees who make significant contributions to the organization					
75	Recognition in form of titles, certificates and commendation letters are periodically awarded to the contributors of knowledge					
76	My organization actively promotes a 'culture of experimentation' that encourages employees to try new ideas and take risks					
77	The mission, vision and objectives of Knowledge Management are clearly propagated throughout POWERGRID					
78	The Enterprise Resource Planning, which is being implemented currently, will provide further boost to knowledge management activities					
79	My organization measures and manages its intellectual capital (IC) in a systematic way					
80	POWERGRID publishes regular intellectual capital reports to its external stakeholders					
81	The evaluation of work/projects that are completed successfully has been done in my organization from KM point of view					

Thank you very much for your help and cooperation

Appendix E

E.1 Executives of Task Force for Implementation of Knowledge Management

S. No.	Name of the department	Name and designation of committee members
1	Information Technology	Ms. Kumud Wadhwa, Chief Manager
2	Corporate Planning	Mr. A.J. Rao, Chief Manager
3	Engineering	Mr. A.K. Singh, Dy. Chief Design Engineer
4	Finance	Ms. Pratyaksha, Chief Manager
5	Operation Services	Mr. V.K. Bhaskar, Chief Manager
6	System Operation	Ms. Minaxi Garg, Chief Manager
7	Human Resource	Mr. K.N. Pandey, Chief Manager

K.N. Pandey, *Paradigms of Knowledge Management*, Studies in Systems, Decision and Control 60, DOI 10.1007/978-81-322-2785-4

Appendix F

F.1 Communities of Practice (CoP) in POWERGRID

S. No.	Community	Anchor	Mobile No.	E-mail
1	O&M	Mr. V.K. Bhaskar	9873549042	vbhaskar@powergridindia.com
2	Telecom	Mr. Vikas Kumar	9560690622	vikas.kr@powergridindia.com
3	Grid Management	Ms. Minaxi Garg	9910378031	mgarg@powergridindia.com
4	Project Management	Mr. G.P. Payasi	9422811688	gppyasi@powergridindia.com
5	Consultancy	Mr. Sunil Samuel	9650894937	sunilsamuel@powergridindia.com
6	Finance	Ms. Pratyaksha	9910378169	pratyaksha@powergridindia.com
7	HR	Mr. Arun Kumar	9910378173	arunk@powergridindia.com
8	IT	Mr. Gautam Srivastava	9431820300	gautam.srivastava@powergridindia.com

K.N. Pandey, *Paradigms of Knowledge Management*, Studies in Systems, Decision and Control 60, DOI 10.1007/978-81-322-2785-4

Bibliography

Asian Development Bank: Disseminating Knowledge Products (2009). google.com. Accessed 27 July 2012

At-Hawamadeh, S.: Knowledge Management: Cultivating Knowledge Professionals. Chandos Publishing, Oxford (2005)

Browne, M.W., Cudeck, R.: Alternative ways of assessing model fit. In: K.A. (1993)

Bukowitz, W.R., William, R.L.: The Knowledge Management Fieldbook. Pearson Education Ltd., Great Britain (1999)

Choi, B., Lee, H.: An empirical investigation of KM Styles and their effect on corporate performance. Inf. Manag. **40**(5), 403 (2003)

Christensen, C.M., Bever, D.V.: SPPTLIGHT: Are investors bad for business. Harvard Bus. Rev. 53–70 (2014)

Conway, S., Sligar, C.: Unlocking Knowledge Asseets: Solutions from Microsoft, Prentice-Hall of India Pvt. Ltd., New Delhi—110 001. P.1 (2003)

da Veus, D.A.: Research Design in Social Research. Sage, London (2001)

Dataware Technologies: Corporate Executive Briefing: Seven Steps to Implementing Knowledge Management in Your Organization; Dataware Technologies inc. www.google.com. Accessed 20 Sept 2010

Denzin, N.: Sociological Methods: A Sourcebook, 2nd edn. Mcgraw-Hill, New York (1998)

Diamantopoulos, A., Siguaw, J.A.: Introducing LISREL. Sage Publications, London (2000)

Duffy, J.: Knowledge management: What every information professional should know. Inf. Manag. J. **34**(3), 10 (2000)

Ein-Dor, P.: Taxonomies, of knowledge. In: Schwartz, D.G. (ed.) Encyclopedia of Knowledge Management, Idea Group Reference, Hershey London, Melbourne, Singapore (2006)

Ellet, W.: The Case Study Handbook: How to Read Discuss and Write Persuasively about Cases. Harvard Business School Press, Boston, Massachusetts, USA (2007)

Fuller, S.: An overview of social epistemology. In: Stehr, N., Grundmann, R. (eds.) Knowledge: Critical Concepts, vol. II, pp. 40–41. Routledge, London and New York (2005)

Faraj, S., Javenpaa, S., Mazchrzak, A.: Knowledge collaboration in online communities. Organ. Sci. **22**(5), 1124–1139 (2011)

Goh, S.C.: Managing effective knowledge transfer: an integrative framework and some practice implications. J. Knowl. Manag. **6**(1), 23–30 (2002)

Hair, J.F. Jr., Black, W.C., Babin, B.J., Anderson, R.E.: Multivariate Data Analysis: A global perspective, 7th edn. Prentice Hall, Englewood Cliffs, NJ (2000)

Hooda, R.P.: Statistics for Business and Economics. McMillan India Ltd., Delhi (2003)

Holsapple, C.W., Joshi, K.D.: A knowledge management ontology. In: Holsapple, C.W. (ed.) Handbook on Knowledge Management, vol. I, pp. 89–90. Springer, New Delhi (2003)

Handzic, M.: Integrated socio-technical knowledge management model: an empirical evaluation. J. Knowl. Manag. **15**(2), 198 (2011)

K.N. Pandey, *Paradigms of Knowledge Management*, Studies in Systems, Decision and Control 60, DOI 10.1007/978-81-322-2785-4

Ipe, M.: Knowledge sharing in organizations: a conceptual framework. Human Res. Dev. Rev. **2**(4), 337–359 (2003)

Jenson, J.K., Matheson, L., Lacey, F.M.: Doing Your Literature Review: Traditional and Systematic Techniques. Sage Publication Ltd., London ECIY IS Colifornia, New Delhi (2011)

Jones, N.B., Mahon, J.F.: Nimble knowledge transfer in high velocity/turbulent environments. J. Knowl. Manag. **16**(5), 724–788 (2012)

Jones, R.G., George, J.M.: Contemporary Management, 7th edn. McGraw Hill Irvin, New York (2011)

Joreskog, K.G., Sorbom, D.: LISREL User's Guide. National Educational Services, Chicago (1981)

Lowler, A.: The role of al basked technology in support of the knowledge management value activities cycle. J. Manag. Inf. Syst. **18**(1), 5–21 (2000)

Linden, T., Cybulski, J.: A hermeneutic study of pattern mining as a knowledge creation process: exploring multimedia design practices. J. Inf. Knowl. Manag **9**(3), 265–266 (2010)

Malik, F.: Managing Performing Living: Effective Management for a New Era. Campus Verlag, Frankfrut/New York (2006)

Mason, D., Pauleen, D.J.: Perceptions of knowledge management: a qualitative analysis. J. Knowl. Manag. **7**(4), 38–48 (2003)

Mason, D.: Five steps in successful knowledge management in your service centre. Customer Relationship Management, Aug 2004 Issue

McLeod, J.: Case Study Research In counseling and Psychotherapy. Sage Publication, New Delhi (2010)

Morokoff, P.J., Quina, K., Harlow, L.L., Whitemire, L., Grimley, D.M., Gibson, P.R., Burkholder, G.J.: Sexual assertiveness scale for women: development and validation. J. Pers. Soc. Psychol. **73**(4), 790–804 (1997)

Newbold, P., Carlson, W.L., Thorne, B.: Statistics for Business and Economics, 6th edn. Pearson Education Inc, India (2008)

Nidumolu, S.R., Subramani, M., Aldrich, A.: Situated learning and the situated knowledge Web: Exploring the ground beneath knowledge management. J. Manag. Inf. Syst. **18**(1), 115 (2001)

O'Dell, C.: The Executive's Role In Knowledge Management. Tata McGraw-Hill Publishing Company Ltd., New Delhi (2004)

Paper Presented at the Annual Spring Meeting of the Psychometric Society in Iowa City (1980)

POWERGRID: Annual report 2013–14 (2015)

Royall, M.D., Bramson, A.L.: Interferance and Intervention: Causal Models for Business Analysis. Routledge, New York and London (2014)

Salisbury, M.: An example of managing the knowledge creation process for a small group. Manag. Lear. **32**(3)

Schwartz, H., Davis, S.: Matching corporate climate and business strategy. Organ. Dyn. **10**(1), 30–48 (1981)

Stanford Encyclopedia of Philosophy–Epistemology: (2015). http://plato.standford.edu/entries/epistemology/

Steiger, J.H., Lind, J.C.: Statistically-based tests for the number of common factors. (1980)

The Theory of Knowledge: www.google.com. Accessed 14 May 2012, p. 1

Van den Hooff, B., Ridder, J.A.: Knowledge sharing in context: the influence of organizational commitment, communication climate and CMC use on knowledge sharing. J. Knowl. Manag. **8**, 117–130 (2004)

Van Mannen, J.: The fact of fiction in organizational ethnography, Administrative. Sci. Q. **24**, 34–41, 539–55

Wickramasinghe, N.: Knowledge creation. In: Schwartz, D.G. (ed.) Encyclopedia of Knowledge Management. Idea Group Reference, Hershey, London (2006)

www.cercind.gov.in. Accessed 12 Dec 2014

www.powergridindia.com. Accessed 12 Dec 2014
Yin, R.K.: Case Study Research Design and Methods, 2nd edn. Sage, Thousand Oaks, CA (1994)
Yin, R.K.: Case Study Research: Design and Methods, 3rd edn. Sage Publication Inc., Thousand Oaks, California (2003)

MIX
Papier aus verantwortungsvollen Quellen
Paper from responsible sources
FSC® C105338

If you have any concerns about our products,
you can contact us on
ProductSafety@springernature.com

In case Publisher is established outside the EU,
the EU authorized representative is:
Springer Nature Customer Service Center GmbH
Europaplatz 3, 69115 Heidelberg, Germany

Printed by Libri Plureos GmbH
in Hamburg, Germany